A PILOT'S METEOROLOGY

A PILOT'S METEOROLOGY

Third Edition

Malcolm W. Cagle

Rear Admiral, United States Navy

and

C. G. Halpine

Late Captain, United States Navy

VNR

VAN NOSTRAND REINHOLD COMPANY

New York / Cincinnati / Toronto / London / Melbourne

Van Nostrand Reinhold Company Regional Offices:
New York   Cincinnati   Chicago   Millbrae   Dallas

Van Nostrand Reinhold Company Foreign Offices:
London   Toronto   Melbourne

Manufactured in the United States of America.

Published by Van Nostrand Reinhold Company
450 West 33rd Street, New York, N.Y. 10001

Published simultaneously in Canada by
Van Nostrand Reinhold, Ltd.

15 14 13 12 11 10 9 8 7 6 5 4 3

*The opinions or assertions in this book are the personal ones of the author and are not
to be construed as official. They do not necessarily reflect the views of either the Navy
Department or the Department of Defense.*

PREFACE

The first edition of *A Pilot's Meteorology* was published in 1941, with a revision which followed in 1952. It is a reflection of the rapid advancement of knowledge that this new revision is necessary. A geometric advancement in man's knowledge of his environment has taken place not only in meteorology but in every field of science.

In the eighteen years since the last revision of this book, the revolution that has taken place in the science of meteorology has been spectacular. The last twenty years have seen the use of weather satellites and upper atmosphere probes; the advancement of the World Meteorological Organization; the establishment of the World Weather Watch; the increased knowledge of high-altitude weather phenomenon; greater knowledge concerning the jet stream; the influence of increased air pollution; the investigation of and the development of embryonic efforts to control weather, rain, and fog; the increase of forecasting facilities for hurricanes, typhoons, severe weather, cyclones, and clear-air turbulence; and the use of computer technology in forecasting. These are some of the many facets of this meteorological revolution. As fast as this field has grown, however, it is certain that the field of meteorology will continue to be vigorous and dynamic as the quest for even greater knowledge, and the ability to apply that knowledge rapidly, continues.

As knowledge in the realm of meteorology and the physics of the atmosphere has increased, so also has the growth in the numbers of pilots and the capabilities of aircraft. Eighteen years ago, light aircraft and operating pilots numbered approximately half of those flying today.

The capability of aircraft has also advanced dramatically in eighteen years—the ability to fly greater distances, cross-country or cross-ocean; the capability to fly at higher altitudes where the weather problems are different; the capability to fly in marginal weather conditions—fog, low visibility, and icing; the development of improved flight instruments, engines, and equipment (such as radar, tacan, heated wings, and nonicing carburetors)—all of these have increased a pilot's capability to fly in marginal weather conditions safely and efficiently.

Hazardous flight conditions rarely develop without some sort of warning. The pilot can anticipate them by making use of the abundant data provided by the weather office and forecaster; by using pilot reports, radar, and radio weather information. Clouds, ice, haze, turbulence, and precipitation are not insurmountable obstacles to successful flight. Flight techniques, instruments, and engines have been developed to overcome such weather hazards.

Thus, this revision has been a major one and much new material has been introduced. Throughout, the author has made an effort to keep paramount the fact that this is a book *by pilots for pilots*. He has sought to confine it, therefore, to the weather problems and phenomena and their solutions which a modern pilot must be prepared to face.

The author has been a naval aviator for almost thirty years and has been involved in the development, education, and supervision of pilots in every aspect of military aviation during this time, including three wars. He did not feel competent, however, to undertake this major revision without expert, experienced meteorological assistance. To this end, he was fortunate to enlist the services of Captain William J. Kotsch, U.S.N., a highly respected meteorologist and author (or co-author) of three meteorological texts and numerous articles. Captain Kotsch is presently the Deputy Commander, U. S. Naval Weather Service Command, and has been a professional member of the American Meteorological Society for many years. His experienced and practical knowledge has been of material assistance in the technical revision throughout this entire text.

I am also indebted to my writer, Chief Yeoman John L. Murray, who labored many months with me to accomplish this task; to Captain S. C. Balmforth, U.S.N., presently with the Naval Weather Service Environmental Detachment, Corpus Christi, Texas, who reviewed and corrected the text; and to Dr. Harner Selvidge and the American Soaring Society for permission to include a condensation of Dr. Selvidge's book as Chapter 20.

Malcolm W. Cagle
Rear Admiral, U. S. Navy

Washington, D. C.
May, 1970

The Publisher would not let this opportunity pass to acknowledge the very substantial contribution to this present book that has been made by the late author of the first and second editions, Charles G. Halpine, Captain, United States Navy, Retired. Suffice it to say here, without his remarkable endeavors, this new edition would have very much less merit: it is not too much to say that the present author and the Publisher "stand on the shoulders of a giant." Our gratitude to Captain Halpine is, indeed, very profound.

CONTENTS

INTRODUCTION

INTRODUCTION

Weather affects and influences the life of every human being, but there are few people in the world more concerned with or more affected by weather than the pilot of an airplane. The farmer looks to the clouds for rain or for fair weather to harvest his crops. The vacationer, the golfer, the picnicker, and the honeymooner look for sunshine and blue skies. The mariner watches wave, wind, sky, and barometer to anticipate his weather. But, if they must, these weather watchers can escape the weather's vicissitudes. Even the sailor at sea, when confronted with a thick fog, a blinding snowstorm, or a gale, can reduce speed or drop anchor and await more favorable conditions.

But to the airplane pilot, weather is a vital factor. He must contend with weather to a far greater extent than people in other professions, and he must share the airspace with clouds, rain, wind, hail each time he flies. He cannot always choose to avoid the weather and sit on the ground. If he is a commercial or military pilot, he must often depart in marginal or poor weather. As he approaches his destination in marginal weather with his fuel running low, he has no choice but to bring his airplane down. Whether the poor weather be fog, icing, or other storm conditions, he must plan for and make a safe landing. The pilot who is well versed in meteorology, who anticipates and recognizes timely weather warnings, who uses the weather "tools" available to him, can avoid many hazardous weather conditions either by altering his route in the proper manner or by landing his plane in safety while there is still time. The pilot who understands the behavior of air currents and other weather conditions is often able to alter his altitude and course and take advantage of tail winds or avoid higher head winds and thereby decrease his flight time and cost of operation. He can learn to avoid ice, turbulence, and thunderstorms.

Unfortunately, a large number of the aviation accidents which occur in the United States are attributable to pilots *without* instrument ratings who stumble into adverse or hazardous weather. In a recent year, there were 4400 general aviation accidents, 420 of which were fatal. In one-third of these fatal accidents the pilots did not obtain sufficient weather information concerning en route and destination weather—information that was readily available. A large percentage of private pilots still take off with an incomplete weather briefing or no briefing at all.

Pilots need not be trained meteorologists, but they should have a working knowledge of the physics of the weather, the thermodynamics of the atmosphere, and an understanding of the meteorological principles which interface with flying in order to be safe pilots—not simply for their own safety, but also for the safety of their passengers and other pilots in adjacent airspace. If accidents are to be reduced, pilots must know and use their weather "tools" to best advantage.

To the military pilot, the ability to fly in any weather is often vital to mission success. In many cases, military operations require flying when prudence would ground even a commercial pilot.

Therefore, all pilots should be solidly grounded in these five fundamentals of weather:

- Direction and velocity of wind at various levels
- Height and thickness of clouds

- En route weather
- Areas and altitudes of icing, severe turbulence, thunderstorms, etc.
- Limited visibility—fog, smoke, precipitation, and low clouds

These five items cover the weather problems which affect flight safety. Professional meteorologists can now forecast such information with reliability. Armed with this, and from his own knowledge and study of weather maps, cloud types and movements, air masses, fronts, and pressure tendencies, the pilot can recognize (with surprising ease) the most important indications of the weather he can expect along his planned route.

The basic principles of meteorology are simple. A new pilot need not be a student of physics and higher mathematics to fly the weather. The theoretical work has already been done for him. Today the weather knowledge that is needed can be learned so easily that there is no excuse for any pilot lacking good practical knowledge of flight conditions.

The purpose of this book, therefore, is *not* to produce professional forecasters or trained meteorologists. Rather, the objective is to present to the pilot, especially to the student pilot, be he private, commercial, or military, basic weather knowledge and fundamentals in order to:

- Promote and enhance air safety
- Receive and use intelligently weather briefing information
- Help student pilots pass their FAA examinations
- Help student military pilots in understanding and passing their weather courses
- Stimulate sufficient interest in all pilots to continue to broaden their weather knowledge by additional reading and research

In Part II, we will present a readable, nontechnical discussion of the physics and thermodynamics of the atmosphere—the composition of the atmosphere, wind and circulation, the formation of clouds, the origins and nature of fronts, and the obstructions to visibility and their causes.

In Part III, we will describe the weather service organizations and meteorological information which are available to the pilot—how to read and understand charts and forecasts. These are the "tools of the trade."

In Part IV, we will give explicit details on *how* to fly the weather; *how* to handle fog, thunderstorms, turbulence, ice; *how* to use radar; *how* to fly at altitude, etc.

In simplest terms, the purpose of this book is to produce safer, better pilots who know and can handle weather.

1 WEATHER IN FLIGHT PLANNING

THE PILOT'S RESPONSIBILITY

It is the responsibility of the pilot to prepare himself and his airplane before each flight, no matter how short or how simple it is expected to be. A major part of this flight preparation concerns weather. To make use of the facilities provided, a pilot must understand what weather information is available, how it is presented, and how it can be used. It is the pilot's responsibility—not the forecaster's or weatherman's—to obtain as complete a picture of the weather at his point of departure, en route, and at his destination as is possible.

To do so, a pilot should tell the forecaster his destination, his estimated time of departure, his planned route, altitude, airspeed, the alternate airfields he can reach, his type of aircraft, and any special information or peculiarities about his airplane that will help the forecaster give a better forecast.

For example, does your airplane have deicing equipment? Radar? Can it fly above the weather? Also, don't hesitate to tell him of your own limitations as a pilot. If you haven't flown in a lot of weather, tell him so. If you are only VRF-rated, be sure and tell him that.

MAKING YOUR OWN ANALYSIS

Weather stations are maintained at all major airports and military airfields for the collection, analysis, and dissemination of weather information essential to flight operations. These stations are staffed by trained observers and forecasters who assist the pilot in his analysis of existing weather conditions and in his interpretation of forecasts.

At other airports, where activity is substantial but less than that of major airports, there is a Flight Service Station (FSS).

At smaller airports which have only limited weather facilities, there is sometimes a no-cost telephone service to the nearest FAA office. Or a pilot may obtain a weather briefing and forecast in certain cities over an unlisted telephone service. These unlisted numbers can be found in the *Airman's Information Manual*.

But since there are more than 8000 airports in the United States, and since the number of private pilots is still growing rapidly, a personal weather briefing is not always possible or available at every airport. Obviously, the private pilot who intends to fly locally in the immediate area of a small, private airfield needs a minimum of weather information. But if your trip is for a distance greater than 250 miles, you should certainly obtain the most complete and detailed weather briefing you can, using all the sources which are available to you. Therefore, a pilot should be capable of making his own weather analysis. Even if he has the *best* meteorological facilities, he should still make his own forecast before he consults or telephones the forecaster. By doing so, he trains himself to understand weather situations, and it helps him understand the forecaster's analysis and to ask sensible questions. (Any pilot who accepts an analysis or forecast which he does not completely understand is negligent.)

To repeat, a pilot's own analysis of the weather will give him a complete picture of the weather conditions and developments that will affect flight along his route and enable him to discuss intelligently any apparent discrepancies in the forecasts. Once in the air, the pilot cannot always consult the forecaster or the maps to see explanations for unexpected changes. He must then rely on his weather knowledge and experience and on the information he obtained before leaving the ground.

PREFLIGHT WEATHER CHECKLIST

In Appendix 4, page 385, you will find a very detailed checkoff list of the weather items you should investigate before you file your flight plan. In this chapter, the following abbreviated checkoff list summarizes the principal weather items you should check:

(1) *VFR or IFR* First of all, determine whether your flight will be conducted under IFR or VFR rules. (To fly Instrument Flight Rules, you must, of course, have an instrument rating.)

(2) *Present Weather at Departure Airfield*
   ■ Surface temperature and pressure altitude
   ■ Surface winds
   ■ Bases and tops of cloud layers
   ■ Visibility
   ■ Precipitation (type and intensity)
   ■ Freezing level
   ■ Climb winds

(3) *Forecast Weather at Departure Airfield*
   ■ What will the weather be in the next 30 minutes—the next hour—at the airfield you are leaving (in case you must return)?

(4) *En Route Weather*
   ■ Bases, tops, type, and amount of each layer of clouds
   ■ Visibility at flight level
   ■ Type, location, intensity, and direction and speed of movement of fronts
   ■ Freezing level
   ■ Temperatures and winds at flight level
   ■ Areas of severe weather (thunderstorms, hail, icing, and turbulence)
   ■ Areas of good weather (in event of an emergency landing en route)

(5) *Forecast Weather for Destination and Alternates*
   ■ Bases, tops, type, and amount of cloud layers
   ■ Visibility
   ■ Weather and obstructions to vision
   ■ Freezing level
   ■ Surface wind speed and direction
   ■ Forecast altimeter setting
   ■ Type and intensity of precipitation, and condition of the landing runway in order to estimate the length of roll-out and braking action (in case of snow and ice)

It should be remembered that if your departure is delayed longer than one and a half hours, it is your responsibility to recheck the weather.

The *forecast* weather at the point of takeoff is often neglected in flight planning. This is important, since knowledge of local weather can be the deciding factor when a decision is required soon after takeoff and you have to return to your point of departure. In some cases, the weather may deteriorate rapidly shortly after takeoff.

Winds up to flight altitudes should be requested in order to compute the distance that will be covered and the fuel that will be consumed during your climb to altitude.

Knowing the types of clouds at flight altitudes en route will give you an idea of the areas of possible precipitation, icing, turbulence, and other hazards to flight. If you get this information *before* you depart, you may be able to make the proper operational decision in the event of an en route emergency.

If yours is a jet or turboprop aircraft, which can fly at high altitudes, it is particularly important that your weather briefing preparation be complete, for the problems of endurance and weather destination accuracy become vital to a safe flight.

GO–NO-GO DECISION

After you have completed your study of the maps, sequences, forecasts, PIREPS, and other advisories, and after discussing the weather with the forecaster, either in person or on the telephone, remember that the go-no-go decision is *yours*. It is *not* the forecaster's.

These procedures seem time-consuming and detailed, but the extra minutes spent in the weather office or on the telephone prior to takeoff may save many hours of uncomfortable or dangerous flying. It is the wise pilot who takes the time to get a good, complete weather briefing!

THE ATMOSPHERE

2 THE ATMOSPHERE

CHARACTERISTICS OF THE ATMOSPHERE

That mass of air which surrounds our planet, covering land and sea alike in an unbroken envelope, is called the *atmosphere* (which means "sphere of vapor"). If the earth were a baseball, the air mass would be about as thick as the baseball's cover. Or, put another way, if the weight of all the air were replaced by the weight of ordinary water, the earth would be covered by water 34 feet deep. Although we cannot see it, the atmosphere is as real and as material as the oceans themselves which cover some three-quarters of the surface of the globe. In some ways, in fact, the atmosphere has the characteristics of an ocean. It has its currents which, when moving horizontally, are called winds. Some of these flow with the regularity of the great currents of the seas, while others are as fickle as the flight of a butterfly. Near the surface of the earth, these currents are retarded or deflected by irregularities of the earth's surface. We find airfalls flowing over the tops of ridges and buildings just as waterfalls plunge over precipices. We find just the reverse, also, for air striking a mountain range may be reflected upward for a considerable distance. The atmosphere becomes less dense and colder, on the average, as we proceed upward.

AIR

Air is the substance of life. We walk in it, we sleep in it, we live and die in it. Air is spread throughout our bodies. It is in the ground beneath our feet. Every day, each one of us breathes 6000 gallons of it.

And, of course, pilots fly in it. It is the substance of the atmosphere, a mechanical mixture of several gases. A sample of "pure" air contains by volume about 78.0% nitrogen, 21.0% oxygen, and 0.9% argon. The remaining 0.1% of air is composed of about 0.03% carbon dioxide and traces of hydrogen and inert gases such as helium, xenon, neon, krypton, and ozone. Air also contains variable amounts of water vapor, most of which is concentrated below 30,000 ft. The maximum amount of water vapor that the air can hold depends on the temperature of the air and the pressure; that is, the higher the temperature, the more water vapor it can hold at a given pressure (but never more than 4% by volume). The average amount of water vapor in the atmosphere is approximately 1.2% by volume.

But completely "pure" air is never found in nature. It always contains variable amounts of impurities such as dust, the ashes of combustion, and minute salt particles. The amount of the impurities is never constant and varies greatly with altitude and location. However, these particles are very important in aviation because of their effect on visibility and especially because of the part they play in the development of clouds, a relationship to be discussed in Chapter 7.

Despite its composition, air conforms to the general laws of gases as if it were composed of only one gas. It is highly compressible and perfectly elastic. Though very light, air has a definite weight which, at ordinary temperature and pressure, is about 1/770 of the weight of water, or about 1.22 oz per cu ft. And at ordinary pressure and temperature a pound of air has a volume of 13.1 cu ft. There are, in fact, about 1.185×10^{19} pounds of air—about 5925 million million tons—bearing down on the earth at all times.

By its weight, air exerts pressure upon everything it touches. At sea level, the average pressure is about 14.7 psi, or about one ton per square foot. This pressure is exerted equally in all directions. Although at sea level the total pressure on a human body amounts to approximately 30 tons, we do not notice it, for it is balanced by an equal pressure from within. Yet it is the same as that exerted by a column of water 33 ft high or by a column of mercury about 30 in. high.

Just as the pressure at any point in a column of water or mercury depends upon the weight of the column above that point, so the pressure of the air depends upon the weight of the air column above the given point. Atmospheric pressure decreases with altitude, slowly approaching zero. (There is no sharp limit at which the atmosphere stops and becomes empty space.) With a normal distribution of temperature, the pressure on a mountaintop is less than that at the bottom of an adjacent valley.

It follows, therefore, that the air in the valley, being under greater pressure, is denser than that on the mountaintop. A cubic foot of air at sea level weighs more than a cubic foot of air at any higher level. Further, because it is easily compressible, a pound of air at sea level occupies less space than a pound of air at any higher point.

In general, as we rise above the surface of the earth, we find that the air grows colder with fair regularity up to an altitude of about six to eight miles. Above that level, the air temperature remains fairly constant up to about 25 miles altitude. From there upward, its temperature increases rapidly again, reaching 350°F. at approximately the 40-mile level. Above that, the increase becomes more gradual as we move toward the outer limits of the atmosphere.

COMPOSITION OF THE ATMOSPHERE

As stated before, air is a mechanical mixture of gases and water vapor. The other gases, however, have no practical bearing on weather. The water vapor, however, is of major importance in meteorology. A molecule of water vapor weighs about five-eighths as much as a molecule of dry air. Without water vapor in the air, we should have few of the phenomena which comprise what we call weather. There would be no clouds to protect us from the heat of the sun, no rain to water the plant life upon which animals and men depend for food. Life in its present form could not exist upon the earth.

It is interesting that, on the average, the water-vapor content of the atmosphere varies with the altitude and the latitude; it is lowest at high altitudes and high latitudes. Although practically all of the water-vapor content is concentrated below the 30,000-ft level, no air with a water-vapor content of zero has been found in nature, just as no "pure" air has been found.

STRUCTURE OF THE ATMOSPHERE

Science divides the atmosphere into three main structural divisions or layers. From lowest to highest they are known, respectively, as the troposphere, the stratosphere, and the ionosphere. (Troposphere comes from the Greek word meaning "turn"; stratosphere from the Latin word meaning "layer" or "covering." Ionosphere gets its name from the fact that it contains vast numbers of electrified air particles or ions.) Since the atmosphere gradually becomes more and more rarefied with increased altitude, there is no definite altitude at which it may be said to end and outer space to begin. In practice, however, the 500-mile point may be taken as the outer limit of our atmospheric

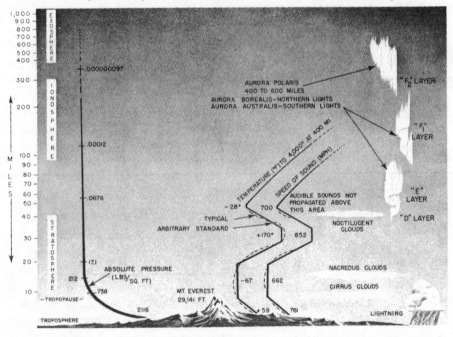

Figure 2–1 Structure and Characteristics of the Atmosphere.

envelope. At that point there are only four molecules per cubic mile, as compared to about 2700×10^{16} molecules per cubic centimeter at sea level.

A common system of identifying the layers is shown in Figure 2–1.

The Troposphere

About one-half of the atmosphere lies below an altitude of 18,000 ft; most of the remainder is concentrated within six to eight miles above the earth's surface. Practically all weather phenomena occur within the troposphere. For this reason, some meteorologists prefer the word "weathersphere" to troposphere. In the troposphere, all six meteorological elements which define weather as the state of the atmosphere are found: air temperature, humidity, clouds, precipitation, atmospheric pressure, and wind velocity. There is considerable motion of winds and convectional currents, which often carry heat and water vapor to high altitudes. The mixing of the air thus effected tends to produce a fairly rapid rate of temperature decrease with altitude.

The troposphere varies in height above the earth from an average of 54,000 ft above sea level over the equator to 29,000 ft over the poles. Its height also varies with seasons; it is higher in the summer than in the winter. The upper limit, where convective currents cease, is called the *tropopause*. This transitional layer is characterized by the fact that temperature is isothermal, that is, remains constant with an increase in altitude. Because of a lack of vertical currents, there are no general weather phenomena within the tropopause.

This book deals mainly with the troposphere since it contains most of the weather affecting aviation, and since the large majority of flying is conducted in it.

Because the atmosphere contains 21% oxygen, the pressure that oxygen exerts is about one-fifth of the total air pressure at any given altitude. This is important to pilots because the rate at which the lungs absorb oxygen depends upon the oxygen pressure. The average person absorbs oxygen at about three pounds of pressure. Since oxygen decreases with altitude, the pilot who makes a prolonged flight at high altitude without oxygen will suffer an impairment of vision and judgment and may even lose consciousness. When the atmospheric pressure falls below 3 psi (about 40,000 ft), breathing pure oxygen is not sufficient and cabin or cockpit pressurization becomes necessary. All high-flying military and commercial aircraft now have pressurized cabins, and they are becoming increasingly common among business and private aircraft.

The Stratosphere

Above the tropopause lies the stratosphere. Of the six elements which define weather, only three are found here: temperature, wind velocity, and pressure. The stratosphere is characterized by the absence of vertical currents. Motion is mainly horizontal in stratified flow, as the name implies.

The base of the stratosphere (the tropopause) does not remain fixed, however. It varies in altitude with the latitude, the season, and the local weather below it. It is lower over the poles than over the equator, lower in winter than in summer, and lower over areas of low pressure than over areas of high pressure. Figure 2–2 shows that the base of the stratosphere averages about 28,000 ft over the pole and rises with fair regularity to above 55,000 ft over the equator.

The stratosphere is almost cloudless, a great advantage to jet air

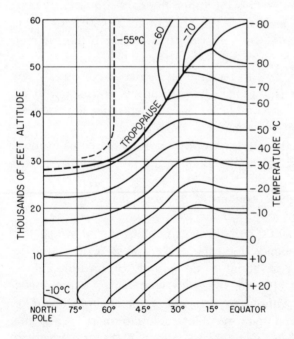

Figure 2–2 Mean Temperature of the Atmosphere, Northern Hemisphere, showing the Variation with Latitude in Temperature Distribution, and Variation in the Altitude of the Tropopause.

travel since it is free from most of the tropospheric storms of the lower altitudes. Occasionally, a special "mother of pearl" cloud, called a nacreous cloud, is observed in connection with the ozone layers.

The Ionosphere

The upper limit of the stratosphere is considered to be about 60 miles above the earth's surface. The range of about 500 miles above this level is known as the ionosphere. The region is so named because there the number of free ions in the atmosphere increases and important electrical phenomena occur. In the winter season, auroras occur throughout the entire region, sometimes extending downward into the upper stratosphere. Ionized layers of the ionosphere have the property of reflecting radio waves, thus greatly affecting their transmission and propagation, an important consideration for pilots.

3 TEMPERATURE

In simplest terms, temperature is the amount of heat or cold measured by a thermometer.

Heat is a form of energy. The energy radiated by the sun (whose surface temperature is about 6000°C.) is one of the primary causes of weather on the earth (whose average temperature is only 20°C.). It is therefore important to understand the effect of the sun's energy on our planet.

First, the sun's energy must pass through the atmosphere which surrounds the earth. (See Figure 3–1.) The atmosphere reflects about 42% of the short-wave solar radiation back into space. About 16% is absorbed by the atmosphere itself near the earth's surface and, therefore, about 43% reaches the earth's surface itself. The earth radiates the heat received from the sun in long waves. This type of radiation is readily absorbed by the atmosphere (primarily by the water vapor in the air). Thus, the earth's

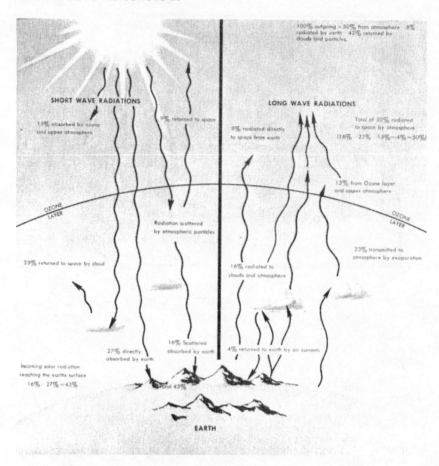

Figure 3–1 How the Earth Receives Its Heat.

atmosphere acts like the glass roof of a greenhouse; it lets the sun's energy through to heat the earth, and in turn, the atmosphere is heated by the earth's radiation.

Three important temperature and heating factors are responsible for weather changes on the earth:

(1) The earth's daily rotation about its axis
(2) Its yearly motion around the sun
(3) The unequal heating received by land and water

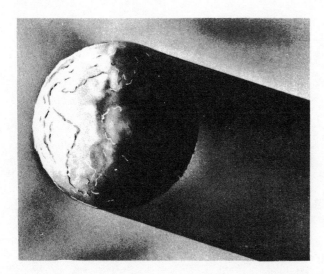

Figure 3–2 Daily Heating and Cooling of the Earth's Surface.

DAILY ROTATION

Daily heating and cooling result from the earth's rotation about its axis. (See Figure 3–2.) As the earth rotates, the side facing the sun is heated; when night comes, this side cools, generally reaching its lowest temperature shortly before sunrise.

YEARLY MOTION

The heating effects due to the earth's yearly rotation around the sun are modified by the "tilt" in the axis of the earth. Areas under the direct or perpendicular rays of the sun receive more heat than those under the slanting rays, as shown in Figure 3–3. The slanting rays pass through more of the atmosphere, which absorbs, reflects, and scatters the sun's energy; therefore, less energy reaches the surface of the earth and the lower atmosphere. (This accounts for the difference in the warmth of sunlight at 0800, when the rays are slanting, and at mid-day, when the sun is nearly overhead.) Each year the perpendicular rays of the sun migrate from 23½° north latitude on June 21 to 23½° south latitude on December 22. It is this migration which causes the seasons of the Northern and Southern Hemispheres. The

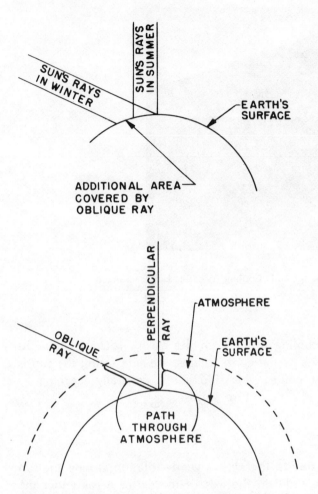

Figure 3–3 Variations in Solar Energy Received by the Earth.

warmest weather of the Northern Hemisphere comes after June 21, when the most heat has been absorbed from the sun, which is nearly overhead each noon. Figure 3–4 shows this migration. Note that the perpendicular rays of the sun strike the earth's surface at 23½° north latitude on June 21, at the equator on September 22, at 23½° south latitude on December 22, and at the equator again on March 21.

Unequal duration of daylight also contributes to the uneven distribution of heat. Figure 3–4 shows that each pole has six months of daylight and six months of night each year. On June 21, all territory

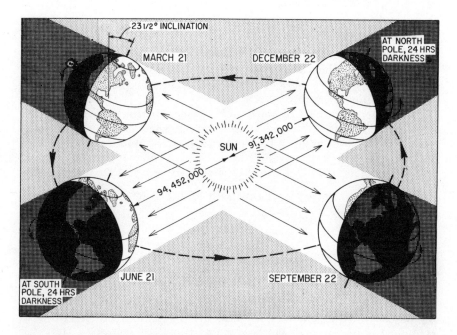

Figure 3–4 Effect of Inclination of the Earth on Seasons.

within the Arctic Circle has 24 hours of daylight; on December 22, all territory within the Arctic Circle has semidarkness or twilight.

LAND MASS AND WATER MASS HEATING

The third temperature factor affecting our weather is the unequal heating received by land and water.

Land areas heat more rapidly than water areas; land also cools more rapidly than water, about three times more rapidly in fact. During the night, water retains its warmth while land loses its heat rapidly to the atmosphere. This difference between land and water also influences seasonal temperatures. In winter, oceanic climates are warmer than continental climates of the same latitude; in summer, the oceanic climates are cooler.

Water absorbs heat more slowly than land surfaces for the following reasons:

(1) Water surfaces reflect about 40% of solar radiation, leaving 60% to be absorbed; most land surfaces reflect only 10 to 15%, absorbing 85 to 90%.

(2) The sun's rays penetrate a deep layer of water, but only a few inches of earth.

(3) The movement of water by wind and current distributes the heat over large areas. The Gulf Stream and the Japanese current, for example, carry heat to the polar regions while the Labrador and California currents carry cold water southward.

(4) The specific heat of water is four times that of land; that is, four times as much heat is required to raise the temperature of a given mass of water as is required to effect the same rise in temperature of an equal mass of land.

(5) Evaporation, which is a cooling process, occurs over the water surface.

However, water retains its warmth longer chiefly because it heats to greater depths.

LOCAL HEAT DISTRIBUTION

In addition to these three main temperature factors which affect weather, there are other local influences: Clouds are excellent reflectors of the sun's heat; they also are a factor in the retention of terrestrial radiation. (See Figure 3–5.) The character of the earth's surface affects the local heat distribution—for example, the color, texture, and amount and type of vegetation influence the rate of heating and cooling. Generally, dry surfaces heat and cool faster than moist surfaces. Plowed fields, sandy beaches, and paved roads become hotter

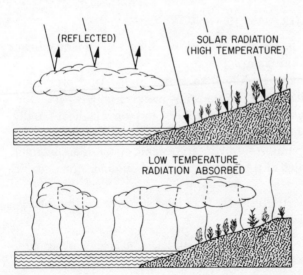

Figure 3–5 Reflection and Absorption of Radiation by Clouds.

than surrounding meadows and wooded areas. During the day, air is warmer over a plowed field than over a forest or swamp; during the night, the reverse is true.

TRANSFER OF HEAT

Pilots should understand the four types of heat transfer: (1) radiation, (2) conduction, (3) convection, and (4) advection. Heat is transferred from the earth to the atmosphere by radiation and conduction, and within the atmosphere by radiation, advection, and convection.

Radiation

Radiation is the process by which energy is transferred through space or through a material medium from one place to another in the form of electromagnetic waves. It is the fundamental process by which the earth is heated; it is also the fundamental process by which the atmosphere is heated (that is, by terrestrial radiation).

Conduction

Conduction is the transfer of heat by contact from molecule to molecule. It is important in meteorology because it is the basic reason that air close to the surface of the earth heats during the day and cools during the night. Although air is a poor conductor (as shown by the use of dead air space in thermopane glass and the air spaces used as insulation in buildings), the heating and cooling of air at the immediate surface of the earth are accomplished by conduction.

Convection

Convection is the vertical transfer of heat by means of vertical currents—movements of warm air to higher levels and the descending of cold air to lower levels. Here, by radiation and conduction, it can absorb heat from the earth. Convection currents are turbulent and cause bumpiness in the air. For example, on a calm, sunny day, air rises over highways and settles over nearby rivers. It is the convection process which makes soaring such a fascinating aerial sport. This will be further discussed in Chapter 20.

Advection

Advection is the transfer of heat by horizontal currents of wind from region to region. If the wind is colder, it is described as cold air advection; similarly, if the wind is warmer, it is described as warm air advection. Of the four processes, advection is probably the most important.

TEMPERATURE

According to the molecular theory of matter, all matter is composed of minute molecules which are in more or less rapid motion. Temperature is a measure of the average velocity of these molecules. As the velocity of molecular motion increases, so temperature increases.

Measuring Temperature

Air temperature is usually measured with a mercury thermometer; special types of thermometers are used for recording and for upper air measurements. (Incidentally, the invention of the thermometer is credited to Galileo in about 1590.)

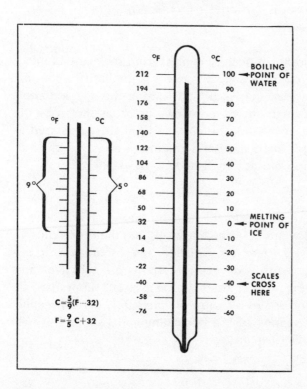

Figure 3–6 Temperature Scales and Conversions.

Temperature Scales

Two fixed temperatures—the melting point of ice and the boiling point of water (at standard pressure)—are used to calibrate thermometers. The two common measures of temperature are Celsius (centigrade) and Fahrenheit. The relationship of the fixed points on each is shown in Figure 3–6. Pilots frequently find it necessary to convert temperature readings from one scale to the other for two reasons: (1) Surface temperatures in the United States and Canada are usually given in the Fahrenheit scale while upper air temperatures are given in the Celsius scale, and (2) many aircraft are equipped with Celsius thermometers. To convert from degrees Celsius to degrees Fahrenheit, use the equation: $°F. = \frac{9}{5} (°C.) + 32$. (Another good thumb rule for conversion is to double the Celsius reading, subtract one-tenth of this product, and add 32 to obtain the Fahrenheit reading.)

Temperature Variation with Altitude

When you get into an airplane and take off, you will notice an overall decrease in temperature as you gain altitude. This is due to the fact that the air nearest the earth is heated the most—it is closer to the radiator. The variation in temperature with altitude is called the *lapse rate* and is expressed in degrees per thousand feet. The *average* lapse rate in the troposphere is about 2°C. per thousand feet, but this average can vary greatly from one day to the next.

The fact that the lapse rate varies is the main reason that temperatures aloft are measured daily. Information important to pilots is determined from the measurement of temperatures aloft, such as:

(1) Levels at which freezing temperatures (icing hazards) occur
(2) Types of clouds that will form
(3) Maximum surface temperatures during the day and minimum surface temperatures during the night (fog formation)
(4) Amount of turbulence in the air

Inversions

An inversion (an inverted or upside down lapse rate) is a temperature change with altitude in which the temperature *increases* for a few thousand feet before the normal temperature decrease begins occurring. Such inversions occur rather frequently, but are generally re-

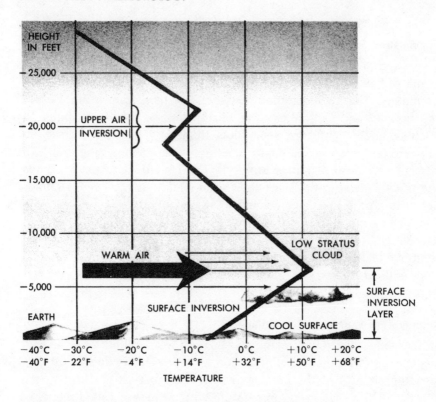

Figure 3–7 Surface and Upper-air Inversions.

stricted to relatively small layers of the atmosphere. Figure 3–7 illustrates how temperature-altitude relationships are represented graphically.

Inversions are formed as follows:

(1) Advection of warm air (a warm wind) brings in temperatures aloft warmer than the temperatures near the surface, causing an inversion aloft.

(2) Vertical descending currents (called subsidence).

(3) A more frequent type of inversion is that produced immediately above the surface of the earth by nocturnal cooling of the earth's surface. The air near the ground is cooled by contact (conduction); this cooling does not extend very high, since air is a poor conductor. At an altitude of about 200 ft, the same air temperature exists at 0600 as existed at 1800 the previous evening. This inversion is shown in Figure 3–8.

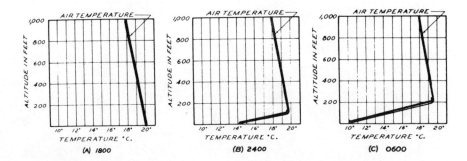

Figure 3–8 A Nocturnal Inversion.

(4) Inversions extending up to 1500 or more feet are possible, owing to nocturnal cooling and adequate wind velocities near the surface causing a mixing of cool surface air with warmer air aloft.

4 PRESSURE

Atmospheric pressure is defined as the force exerted by the weight of the atmosphere on a unit area. Knowledge of atmospheric pressure is most important to pilots in determining altitude, understanding the functioning of altimeters, and forecasting weather.

PRESSURE INSTRUMENTS

The instruments commonly used in the measurement of atmospheric pressure are (1) the mercury barometer, (2) the aneroid barometer, and (3) the barograph.

Mercury Barometer

The mercury barometer was invented in 1643 by the Italian Torricelli. It is a graduated glass tube of uniform internal diameter, somewhat longer than 30 in., which is closed at one end and completely filled with pure mercury. (Mercury is used because it is one of the heaviest liquids at normal temperatures.) The open end of the mercury tube is then temporarily inverted, sealed and placed down in a cup which also contains mercury. When the temporary seal over the open end is removed, the mercury in the tube falls until the top of the column is about 30 in. above the level of the mercury in the cup, leaving a vacuum in the upper or closed part of the tube. Since the weight of the column left standing in the tube is exactly balanced by the pressure which holds it up—namely, the atmospheric pressure—it follows that the height of the column will vary as the supporting pressure varies. As the atmospheric pressure decreases, the height of the supported mercury column will be decreased; as the pressure increases, the mercury column will rise higher. Each change in height of the column will be directly proportional to the change in pressure. Since the mean pressure is nearly 15 psi, and the mean barometric height (at sea level) is about 30 in., a change of

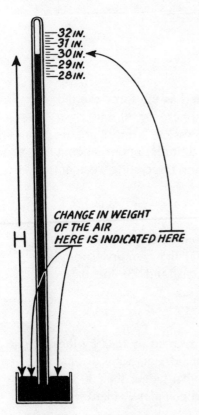

Figure 4-1   Mercury Barometer.

(A)

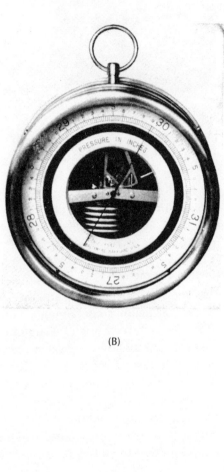

(B)

Figure 4–2 (A) Mercury Barometer. (B) Aneroid Barometer. (Courtesy Bendix-Friez.)

Figure 4–3   Altimeter with Kollsman Window. (Courtesy Kollsman Instrument Corporation.)

1 psi in pressure is, therefore, represented by a 2-in. change in the height of the mercury column.

Figure 4–1 shows the principle of the mercury barometer. The height H depends upon the atmospheric pressure; pressure may then be measured by the height of the mercury column. The height H, measured in inches, will give the barometric pressure in inches of mercury.

Aneroid Barometer

The aneroid barometer ("aneroid" means "without liquid") consists of a metal bellows containing a partial vacuum. The bellows expands or contracts in response to changes in atmospheric pressure. A pointer, linked to the bellows, moves across a calibrated dial, thereby providing the indicator mechanism. (See Figure 4–2B.) Although liable to certain types of minor errors, and not as accurate as the mercury barometer, the aneroid barometer is useful because of its compact and rugged construction and its light weight. The aircraft altimeter (Figure 4–3) is nothing more than an aneroid barometer calibrated to read altitude directly in feet rather than units of pressure.

Barograph

The barograph is an aneroid barometer which gives a continuous record of atmospheric pressure. (See Figure 4–4.)

UNITS OF PRESSURE MEASUREMENT

While there are several systems for measuring pressure, only two are of interest to pilots.

Figure 4–4 Microbarograph. (Courtesy Environmental Science Division, The Bendix Corporation.)

Inches of Mercury

The most familiar measurement unit is inches of mercury, derived from the height of the mercury column in a mercury barometer. Since the inch is a unit of length, this system does not express directly the force per unit area that the atmosphere exerts.

Millibar

In meteorology, it is more convenient to express the pressure directly as a force per unit area. This was done, commencing in 1939, by introducing a unit called the *millibar* (mb). This unit, as the name suggests, is one-thousandth of a bar. The *bar* is a unit of pressure which represents a force of one million dynes per square centimeter. (A dyne is the metric unit of force required to give a mass of 1 gram an acceleration of 1 centimeter per second per second.) When expressed in millibars, the atmospheric pressure is given directly as a force per unit area. Normal atmospheric pressure is 1013.2 mb, which is equivalent to 29.92 in. of mercury.

SEA-LEVEL PRESSURE

The altitudes of mountains and the depths of oceans are measured from mean sea level. Mean sea level is also used as the reference level denoting the bottom of the atmosphere, and mean sea level atmospheric pressure is measured on this basis.

The "standard" atmospheric pressure at mean sea level at a standard air temperature of 15°C. is 1013.2 mb, 29.92 in. of mercury, or 14.7 psi. However, actual sea-level pressures in the atmosphere can vary from as low as 950 mb (about 28 in.) to as high as 1050 mb (about 31 in.).

STATION PRESSURE

The atmospheric pressure computed for the level of the station elevation at an airfield at a given time is called the *station pressure*. This may or may not be the same as the actual pressure, the difference being attributable to the difference in reference elevations. The station pressure is dependent on the altitude of the airfield, the effect of gravity, and the amount of air above the station.

It is important for pilots to understand station pressures since they have an effect on takeoff and landing speeds, stall speeds, and rate of climb. For example, the average light plane taking off from Miami (at sea level) might require only a 1000-ft run. But at Denver (5300 ft above sea level) the takeoff run might be 1800 ft.

PRESSURE TENDENCY

The pressure tendency (or barometric tendency) is the type and amount of change in the atmospheric pressure during a given period of time. This is of great importance to the pilot as well as to the meteorologist. Pressure tendency plays a major part in weather forecasting and is regularly reported with other observed data. The unit used is the total barometric change occurring in the three-hour period immediately preceding the reported observation expressed in millibars and tenths. (See Chapter 12, page 197.) Put in everyday terms, a "falling" barometer usually indicates that storm clouds, rain, or snow is coming, while a "rising" barometer usually indicates fair weather.

ATMOSPHERIC PRESSURE AND ALTITUDE

Atmospheric pressure decreases as altitude increases because the pressure at a given point is a measure of the weight of the column of air above that point. As altitude increases, the pressure is diminished by the weight of the air column below. Since air is a compressible fluid, the lower layers are, for the same thickness, heavier than the higher layers. Consequently, the pressure decreases less and less slowly with altitude. (See Table 4–1.) Cold, heavy air masses have a more rapid rate of decrease of pressure with altitude than warm air masses.

MAPPING PRESSURE AREAS

A surface weather map has sea-level pressure plotted in millibars for each reporting station, regardless of the altitude of the station. A line can be drawn connecting equal values of this reported pressure. Such a line is called an *isobar*. Isobars outline pressure areas in somewhat the same manner as contour lines outline terrain features on contour maps.

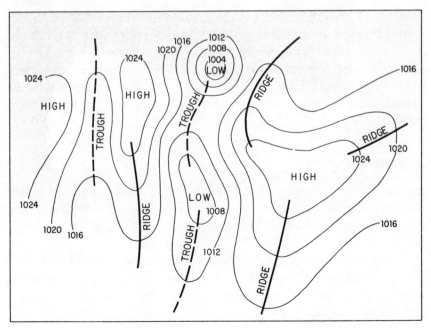

Figure 4–5 Pressure Systems.

PRESSURE SYSTEMS

Isobars on a weather map show some areas of relatively high pressure and other areas of relatively low pressure. Common types of pressure systems, sometimes called pressure patterns, are illustrated in Figure 4–5. The five types of pressure areas are:

(1) *Low* A low is a region of the atmosphere where the barometric pressure is lower than the surrounding pressure, usually encompassed by closed isobars with the point of minimum pressure as its center.

(2) *High* A high is a region of the atmosphere where the barometric pressure is higher than the surrounding pressure, usually encompassed by closed isobars with the point of maximum pressure as its center.

(3) *Col* A col is a saddleback region between two highs or two lows.

(4) *Trough* A trough is an elongated area of low pressure, with the lowest pressure along the trough line.

(5) *Ridge* A ridge is an elongated area of high pressure with the highest pressure along the ridge line.

PRESSURE GRADIENT

The rate of change in pressure in a direction perpendicular to the isobars is called the pressure gradient. The rate of change in height in a direction perpendicular to the contour lines is called the contour or height gradient. The gradient is said to be steep or strong when isobars or contours are close together; and flat or weak when the isobars or contours are far apart. Figure 4–6 shows weak and strong pressure gradients.

PRESSURE ALTIMETER

As stated above, the pressure altimeter is an aneroid barometer graduated to indicate altitude in feet instead of units of pressure. Pilots should always remember that an altimeter really measures *pressure*, not *altitude*.

There are four altimeter adjustments or errors, two of which all pilots should be familiar with: (1) an adjustment or error caused by nonstandard *pressure*; (2) an adjustment or error caused by nonstandard *temperature* aloft. These will be discussed below. The other two errors are gravity error and instrument error.

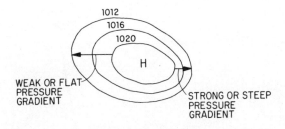

Figure 4–6   Pressure Gradient.

Altimeter Adjustment for Nonstandard Pressure

Because of variations in pressure at sea level, altimeters are designed so that an adjustment can be made to correct for nonstandard surface pressure.

Under most operating conditions, the standard procedure is to set the altimeter to read zero at sea level. During flight, the altimeter will read the altitude above sea level.

Pressure is usually different at the point of landing from that at takeoff (and therein lies a major source of altimeter error). Even though the altimeter is correctly set at takeoff, it can be considerably off at time of landing, even at the same airfield. A pilot should always remember that a change of .30 in. in the altimeter setting will change the height reading about 300 ft, and that on some flights this amount of barometric change can occur during a 200-mile flight.

In landing under conditions of poor visibility or low ceiling, it is vital that the altimeter be set to read the correct altitude. Correct altimeter settings can be obtained by radio from the tower where you are landing. If the field where you are landing does not have radio facilities, you should obtain the expected altimeter setting before you take off. If no accurate setting is available, a general idea of the existing pressure system will be helpful.

Figure 4–7 shows the pattern of isobars in a cross section of the atmosphere from New Orleans, Louisiana, to Miami, Florida. The pressure at New Orleans is 1009 mb and the pressure at Miami is 1019 mb, a difference of 10 mb. An airplane takes off from Miami to fly to New Orleans at an altitude of 500 ft. However, due to a decrease of 10 mb from Miami to New Orleans, it gradually loses altitude, and although the altimeter still reads 500 ft, the aircraft will actually be flying at approximately 200 ft over New Orleans. The correct altitude can be

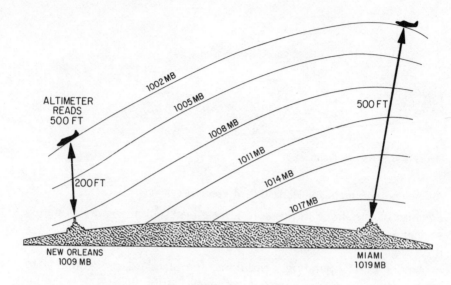

Figure 4–7   Altimeter Error Due to Change in Surface Pressure.

found by obtaining the correct altimeter setting from New Orleans and resetting the altimeter.

The usual procedure is to adjust the altimeter to the local airfield setting just prior to takeoff. (If this setting is not available, you can set the instrument by rotating the scale so that the indicated altitude is equal to the elevation of the airport.) The "setting" indicated in the small window (called the "Kollsman" window) of the altimeter will then be the proper altimeter setting. (See Figure 4–3.) This can be done only when on the ground. (At and above 18,000 ft, all altimeters must be set to 29.92 in.) If you are above 18,000 and descending, FAA rules require that you adjust your altimeter to the local setting provided by ground control.

Here are two useful pilot thumb rules which are accurate up to about 5000 ft: (1) 34 mb = 1 in. mercury = 1000 ft elevation, and (2) a change of 10 mb (which is common) results in an altimeter error of about 300 ft. (See Table 4–1.)

Error Due to Nonstandard Lapse Rate

The second altimeter error is due to nonstandard temperature aloft. Even though an altimeter is properly set for surface conditions, it will usually not be correct at higher levels. Note in Figure 4–8 that, if the

Table 4–1   Atmospheric Pressure Averages with Altitude

HEIGHT (FEET)	PRESSURE (PSI)	PRESSURE (INCHES OF MERCURY)	PRESSURE (MILLIBARS)
50,000	1.68	3.44	116.0
45,000	2.14	4.36	147.5
40,000	2.72	5.54	187.6
35,000	3.46	7.04	238.4
30,000	4.36	8.88	300.9
25,000	5.45	11.10	376.0
20,000	6.75	13.75	465.6
18,000	7.34	14.94	506.0
16,000	7.97	16.21	549.1
14,000	8.63	17.57	595.2
12,000	9.35	19.03	644.4
10,000	10.11	20.58	696.8
8,000	10.92	22.22	752.6
6,000	11.08	23.98	812.0
4,000	12.69	25.84	875.1
2,000	13.66	27.82	942.1
0	14.70	29.92	1013.2

air is warm, the indicated altitude will be less than the true altitude. If the air is cold, the indicated altitude will be greater than the true altitude. Many crashes have resulted because pilots flying by instrument in cold weather did not understand this altimeter error and did

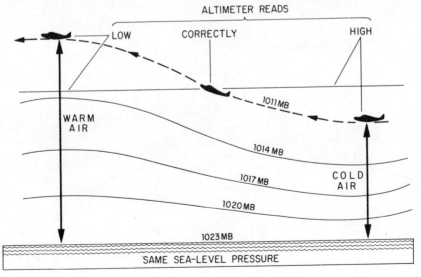

Figure 4–8   Altimeter Error Due to Nonstandard Air Temperatures.

not allow a large enough safety margin to clear mountains. Correction for nonstandard temperature is particularly important during flight over mountains, as there is the additional effect of a local pressure drop in their vicinity.

5 MOISTURE

One of the most important factors of the atmosphere is moisture, which is found in three states: solid, liquid, and gaseous. As a solid, it takes the form of snow, hail, sleet, frost, ice-crystal clouds, and ice-crystal fog. As a liquid, pilots encounter it as dew, drizzle, rain, and as minute particles in the form of fog. In the gaseous state, water forms an invisible vapor—and water vapor is the basis for all clouds. Pilots are also concerned when moisture changes from one state to another—liquid to solid—rain to ice, for example.

As was stated in Chapter 2, most of the atmosphere's moisture is found in the lower troposphere, below 30,000 ft, and its primary source is the oceans (mainly the tropical oceans), although the atmosphere also gains moisture from lakes, rivers, swamps, snow, ice, and even vegetation. Some moisture gets whipped into the air from oceans and lakes by strong surface winds, but most of it gets into the atmosphere by evaporation.

Therefore, practically all weather that interferes with the operation of aircraft is directly associated with water vapor in some form. A clear understanding of moisture is essential to the study of weather, particularly in regard to one of the most serious hazards to aviation, icing of aircraft.

CHANGES OF STATE—EVAPORATION, CONDENSATION, AND SUBLIMATION

Although these terms have been used in establishing the methods of humidity measurement, it is essential to consider a more detailed explanation of these changes of state as they occur in the atmosphere.

Evaporation

Water undergoes the process of evaporation when it changes from the liquid to the gaseous state. As has been stated, all matter consists of molecules in motion. The molecules in a bottle of liquid are restricted in their motion by the walls of the bottle. But on a free liquid surface exposed to the atmosphere, the motion of molecules in the liquid is restricted only by the weight of the atmosphere or, more precisely, by the atmospheric pressure. In effect, molecules escape from the surface of the liquid and enter the air as water vapor. As the temperature of the water is increased, the speed of the molecules is increased, and the rate at which the molecules escape from the water also increases.

During the process of evaporation, heat is absorbed; the amount absorbed is approximately 540 calories per gram of water at a temperature of 100°C. (The calorie is a unit of heat energy; it is the heat required to raise the temperature of 1 g of water 1°C. under specified conditions.) This energy is required to keep the molecules in the vapor state and is called the latent heat of vaporization. Thus, evaporation is a cooling process.

Condensation

Basically, condensation is the opposite of evaporation, in that water vapor undergoes a change in state from gas to liquid. However, a condition of saturation (which means "filled up") must exist before condensation will occur.

Consider a vessel partially filled with water and having a stationary column of air above it. The molecules escape into the air and continue their motion as water vapor. These vapor molecules, because of their motion, exert a pressure, called the partial pressure of the water vapor. The partial pressure of the water vapor is independent of the partial pressures due to the other gases in the air. The rate at which the molecules leave the surface of the water depends only on the temperature of the water. The rate at which the molecules enter the water from the air depends entirely upon the frequency with which they strike the surface, which in turn depends upon the number and velocity of the molecules (the partial pressure of the water vapor in the air). As more and more molecules enter the vapor state, some, in their motion, will return to the liquid. When a condition of equilibrium is reached wherein the number of molecules leaving the liquid equals the number returning, the space above the liquid is said to be *saturated,* and the pressure exerted by the water vapor molecules is called the saturation vapor pressure for the particular temperature existing. When the temperature is raised, the number of molecules necessary for saturation, and their individual velocities, are increased, thus raising the saturation vapor pressure.

In the process of condensation, the heat that was absorbed in evaporation is released from the water vapor into the air and is called the latent heat of condensation. Condensation has a heating effect; more strictly, it lessens the cooling effects of other processes.

Sublimation

The direct change of state from solid to gas and vice versa, without passing through the intermediate liquid state, is called sublimation. This process takes place at temperatures below 0°C. and is similar to evaporation and condensation in that heat is liberated when solidification takes place and is absorbed when vaporization occurs.

At temperatures below 0°C., ice or snow can sublime directly into the air as water vapor. Under similar temperature conditions, when the air has become saturated with water vapor, and the required nuclei are present, the water vapor will sublime directly into ice crystals.

Figure 5–1 illustrates the heat exchanges involved in the processes of evaporation, condensation, and sublimation.

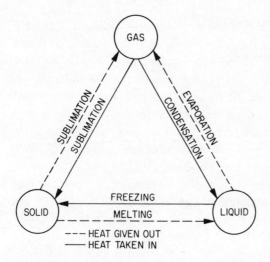

Figure 5–1 Changes in State.

SATURATION

In the atmosphere, saturation occurs by:

(1) Evaporation into the air from a free water surface (ocean, lake, river, raindrops) until vapor pressure equilibrium between the water and the air is reached.

(2) Lowering of the air temperature, thereby decreasing the saturation vapor pressure until it is the same as the actual vapor pressure of the water vapor in the air.

As saturation in the atmosphere is approached, the presence of nuclei, such as dust and salt particles, is required to promote the actual condensation of the water vapor into liquid.

As we noted in Chapter 2, water vapor is a universal part of the atmosphere. If all the water vapor in the atmosphere were to fall in a single shower, it would cover the earth with a layer of water about one inch deep. Notwithstanding, there is a limit to the quantity of water vapor which can be contained in a given volume of air. For example, if more and more water vapor is forced into a closed box, a point will be reached when water will either condense as fog within the box or collect as dew on its walls. If still more water vapor is added, more will condense, and the total amount of vapor in the box will remain unchanged. When this happens, as stated earlier, the volume of the box is said to be saturated with water vapor.

The quantity of water vapor in the saturated volume depends on the temperature. The higher the temperature of the air (and water vapor), the greater is the tendency for liquid water to turn into vapor. At a higher temperature, therefore, more vapor must be injected into the given volume before the saturated state will be reached and dew or fog will form. Conversely, cooling the saturated box will force some of the vapor to condense, and the quantity of vapor in the volume will diminish.

DEW POINT

The dew point is that temperature to which the air must be cooled at constant pressure and moisture content to bring about a state of saturation. Suppose the air temperature is 60°F. and the dew point is 50°F. We should have to cool the air to 50°F. to make it saturated. If the air is further cooled to 49°F., it could no longer hold all the water vapor represented by the former 50°F. dew point—some of it would be virtually "squeezed out" in the form of water; in other words, condensation would have taken place. If this cooling took place only in the fraction of an inch or so of the air above the ground, we should have dew formed; however, what is more likely and quite common is that the cooling will affect a fairly large portion of the low troposphere and we shall have fog.

Hence, the dew point is important to the pilot because it indicates the temperature at which fog will start to form when the temperature is falling. When the air is saturated, we say the relative humidity is 100%; if the air is not saturated, the temperature is a higher value than the dew point and the relative humidity is less than 100%. Pilots must always remember that the greater the "spread" between the temperature, T, and the dew point, T_d, the more remote is the danger of fog formation. The dew point is included in Aviation Weather (Sequence) Reports because it is a critical temperature, indicating the behavior of water in the atmosphere. When the surface air temperature is higher than the dew point, and the spread is increasing, any existing fog and low clouds are likely to dissipate because the air is capable of holding more water vapor. This is especially true in the morning hours when air temperature near the ground is increasing. On the other hand, pilots should be alert for the possibility of fog or low cloud formation at any time when the surface air temperature is within 4°F. of the dew point, and the "spread" is decreasing.

SPECIFIC HUMIDITY

When air rises, it expands and there will be less of it in any given volume. It is convenient to choose a way of measuring the moisture content of the air which will not change as the air moves. This can be done by expressing the humidity as the amount of water vapor, in grams, contained in one kilogram of natural air. This is called the *specific humidity*. A kilogram of air will carry the same number of grams of water vapor with it wherever it goes, unless some is added by evaporation of more water vapor into the air or removed by cooling and condensation of some of the moisture in the air.

Why is familiarity with specific humidity necessary for the pilot? He is never going to be asked to calculate the weight of water in a kilogram of air. The answer is that it is mandatory that he be able to identify and recognize the caliber of specific humidity (high or low), because the higher the moisture content of any portion of the troposphere, the greater will be the quantity of weather (clouds, precipitation, etc.) in it when condensation occurs. A pilot can measure the amount of moisture in the air by watching the dew point temperature (T_d) given on the sequence reports. The higher the dew point, the greater the amount of moisture in the air.

An important rule is that in the normal range of temperatures experienced over the surface of the earth, a doubling of the moisture content of the air is represented by approximately a 20°F. rise in the dew point (see Figure 5–2). If, for example, a kilogram of air with a dew point of 60°F. contains 11 grams of moisture (water vapor), a kilogram of air with a dew point of 80°F. will contain about 22 grams of water vapor. Conversely, if a saturated parcel of air at 80°F. is cooled down to 60°F., 11 grams of water would have to be released through the condensation process. Remember that this condensation is giving off heat energy to the atmosphere and this heat energy will be realized as weather; the turbulence in a thunderstorm and the destructive forces within a hurricane are fueled by this heat energy. In subsequent chapters we will more ably appreciate this knowledge.

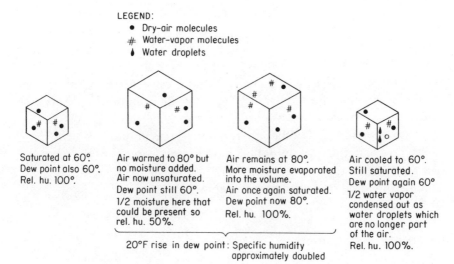

LEGEND:
- • Dry-air molecules
- # Water-vapor molecules
- ♦ Water droplets

Saturated at 60°.
Dew point also 60°.
Rel. hu. 100°.

Air warmed to 80° but
no moisture added.
Air now unsaturated.
Dew point still 60°.
1/2 moisture here that
could be present so
rel. hu. 50%.

Air remains at 80°.
More moisture evaporated
into the volume.
Air once again saturated.
Dew point now 80°.
Rel. hu. 100%.

Air cooled to 60°.
Still saturated.
Dew point again 60°
1/2 water vapor
condensed out as
water droplets which
are no longer part
of the air.
Rel. hu. 100%.

20°F rise in dew point: Specific humidity
approximately doubled

Figure 5–2   Diagrammatic Illustration of Dew Point.

RELATIVE HUMIDITY

The major portion of the atmosphere is not saturated. It contains less than the quantity of water vapor necessary to saturate it. For weather analysis, it is desirable to be able to say how near it is to being saturated. This is expressed as relative humidity (R.H.), which is defined as the ratio of the amount of water vapor contained in the air to the amount of water vapor that the air would contain when saturated at the same temperature. It is expressed as a percentage, saturated air having 100% relative humidity. It is apparent that this ratio is an equivalent factor—relative to the temperature, T, and the dew point temperature, T_d. The temperature of the dew point is in direct proportion to the actual moisture content of the air—actual vapor pressure. The temperature of the air implies to us that if T_d were exactly equal to it in value, the air would be saturated at that temperature; it also implies the amount of moisture the air could contain at that temperature—saturation vapor pressure.

$$R.H. = 100 \times \frac{\text{vapor pressure}}{\text{saturation vapor pressure}}$$

For example, in Figure 5–2 consider an air parcel with T of 80°F.,

T_d of 60°F., and R.H. of 50%. This parcel of air has a particular vapor pressure corresponding to $T = 80$°F.; however, it can hold twice as much water vapor, since the relative humidity is 50%. We can reach an R.H. of 100% by evaporating more water vapor into the air until T_d has been raised to 80°F.; we could also have reached an R.H. of .100% by cooling the temperature to 60°F. Recalling our recently expressed relationship between specific humidity and the dew point, we can recognize that, although a saturated parcel of air at 60°F. contains one-half the moisture that a saturated parcel of air of 80°F. contains, both the parcels have 100% relative humidity—a relative value depending upon the percentage of saturation.

HUMIDITY INSTRUMENTS

Hair Hygrometer

Relative humidity is usually measured with a hair hygrometer ("hygro" means moisture), which operates on the principle that oil-free human hair stretches upon being wet, the stretch being proportional to the degree of saturation of the air with water vapor, or to its relative humidity. The hair is mounted under tension to operate a dial which indicates (hygrometer), or a pen which records (hygrograph) the relative humidity directly. Such an instrument is subject to large errors and is slow to respond to changes.

Sling Psychrometer

More satisfactory, except at temperatures below freezing, is the sling psychrometer. (See Figure 5–3.) This instrument consists of two thermometers, one of which is kept wet by a linen wick provided for that purpose. The cooling of the wet-bulb thermometer is proportional to the rate of evaporation of the water from it, which, in turn, depends on the relative humidity and the temperature of the surrounding air. When the air is very dry, the evaporation from the wick will cool the wet-bulb thermometer, and its temperature indication will be lower than that of the dry bulb. When the air is saturated, there is no evaporation and both thermometers read the same. Psychrometric tables are provided which enable one to obtain dew point and relative humidity from the air temperature given by the dry-bulb thermometer and the temperature given by the wet-bulb thermometer.

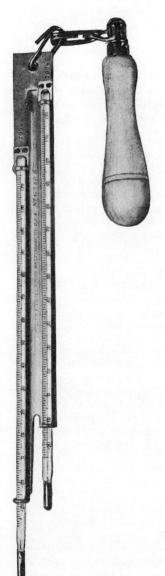

Figure 5–3 Swivel Sling Psychrometer, Weather Bureau Type. With a wet wick attached to one thermometer bulb, the instrument is whirled around by hand, producing evaporation at the wet bulb. From the wet and dry bulb reading the relative humidity is obtained from the humidity tables. (Courtesy Bendix-Friez.)

FORMS OF PRECIPITATION

Causes

Precipitation is the general term for all forms of falling moisture which, more specifically, includes rain, snow, hail, sleet, and their modifications. The chain of events that lead up to precipitation is as follows:

(1) Saturation of the air by:
    (a) cooling and/or
    (b) increasing the weight of water vapor in the air.
(2) Condensation of the water vapor through the action of hygroscopic nuclei, forming clouds. Clouds may be composed of liquid or solid forms of water, or combinations of both, depending upon the temperature in the cloud and the amount of convective activity. There are:
    (a) water clouds,
    (b) water and ice crystal clouds,
    (c) ice crystal clouds.
(3) The precipitation process clouds may be considered as colloidal suspensions of water in the air. The tendency for cloud particles to coalesce and produce larger droplets which fall because of gravitational attraction is promoted by the following conditions:
    (a) nonuniform electrical charges on cloud droplets (electrical attraction),
    (b) differences in sizes of cloud droplets (mass attraction),
    (c) temperature differences between cloud droplets (vapor pressure differences),
    (d) motion of the droplets (turbulent mixing),
    (e) ice crystals (water droplets will evaporate and sublime on ice crystals, increasing their size).

Combinations of these five conditions will determine the size of precipitation particles falling from the clouds. In highly turbulent clouds which extend above the freezing level, both ice crystals and water droplets are present, causing rapid transfer of water from droplet to crystal through evaporation and sublimation. As the ice crystals grow at the expense of the water droplets, they will fall through the cloud, continuing to grow in size until they leave the cloud.

Liquid Precipitation

(1) *Rain*—precipitation which reaches the earth's surface as water droplets. It is classified as light, moderate, or heavy.

(2) *Drizzle*—precipitation from stratiform clouds in the form of numerous drops of water, much smaller in diameter than those occurring in rain. The fact that these minute drops reach the earth indicates the absence of turbulence. Drizzle also is classified as light, moderate, or heavy.

Freezing Precipitation

(1) *Freezing Rain*—precipitation in the form of rain, a portion of which freezes and forms a smooth coating of ice upon striking exposed objects (the wing of an airplane in flight or the surface of a runway, for instance).

(2) *Freezing drizzle*—precipitation in the form of drizzle which freezes in a manner similar to freezing rain.

DEW AND FROST

During clear, still nights, vegetation often cools by radiation to a temperature at or below the dew point of the adjacent air. Moisture then collects on the leaves just as it does on a pitcher of ice water in a warm room. Heavy dew is often observed on grass and plants when there is none on sidewalks or large solid objects. The latter absorb so much heat during the day or give up heat so slowly that they do not cool below the dew point of the surrounding air during the night.

Dew does not "fall." The moisture comes from the air in direct contact with the cool surface. When the temperature of the collecting surface is at or below the dew point of the adjacent air, and the dew point of the adjacent air is below the freezing point, frost will form. Sometimes dew forms and later freezes, but frozen dew is transparent whereas frost is opaque. As will be discussed in Chapter 16, frost can be a dangerous condition on the wings of an airplane.

6 WIND AND ATMOSPHERIC CIRCULATION

INTRODUCTION

Wind is moving air, and while normally referred to as "wind," in certain locations local wind phenomena have acquired exotic names. The Mediterranean has its *mistral*, South America has its *pampero* and *williwaw*, the Rockies of the United States have their *chinook* or *foehn*, Southern California has its *Santa Ana*, Spain has its *solano*, Egypt its *khamsin*, and Southeast Asia its *monsoon*. And there are many more. (See Glossary, Appendix I, p. 345.)

Whatever it is called, a pilot is concerned with wind on every flight. Winds near the surface are important to him in landing, takeoff, and in flight. At upper levels a knowledge of wind phenomena is also essential for navigation, fuel management, and safety.

This chapter will discuss the causes of wind, both winds aloft and surface winds, and the general circulation of the atmosphere.

CAUSES OF WIND

There are several factors which cause wind and the circulation of the earth's blanket of air, and the weather which results from this circulation. But in simplest terms wind is caused by three things: (1) different regions of pressure within the atmosphere, caused by unequal heating of the atmosphere; (2) the earth's rotation; and (3) friction.

The process begins with the sun heating the earth's atmosphere. The air at the equator, being nearest to the sun and receiving the sun's direct rays, becomes warm and rises, flowing generally toward the North and South Poles. (See Figure 6–1.) In the troposphere, beneath this rising equatorial and warmer air, a region of low air pressure is thus created—due to the loss of air—while the air farther north and south of the equator is at *higher* pressure due to *lower* temperatures.

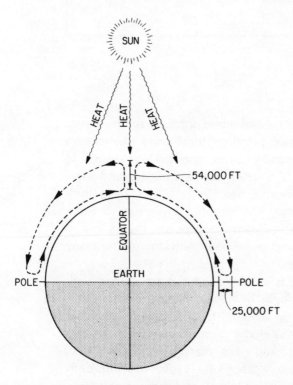

Figure 6–1 Simple Circulation. If the earth stood still, and if it were covered by ocean only, the circulation of the air would be as shown.

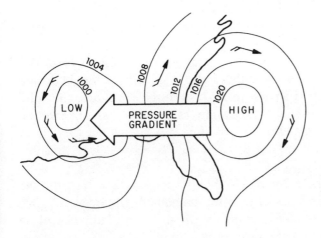

Figure 6–2 Pressure Gradient and Wind Flow Pattern.

This colder, higher pressure air tends to flow toward the equator. Thus, a circulation is established.

The intensity of the force acting from high pressure toward low pressure is indicated by the spacing or distance between isobars— called the *pressure gradient* as shown in Figure 6–2. The closer the isobars are, the steeper the pressure gradient, the greater the pressure force, and the stronger the wind. The pressure gradient always points directly across the isobars toward lower pressure.

Air, therefore, moves from a region of *higher* pressure to a region of *lower* pressure, and the strength of the wind is determined primarily by the pressure gradient, with strong gradients causing strong winds.

THE CORIOLIS FORCE

As stated above, air tends to flow in a direct path from areas of higher pressure toward areas of lower pressure—in other words, air follows the direction of the decreasing barometric pressure. The earth's rotation, however, deflects this flow. This deflecting force is called the *Coriolis force,* so named because it was first described mathematically by the French scientist Gaspard de Coriolis in 1835. (It was also demonstrated in 1856 by an American meteorologist, William Ferrel, who formulated a law which goes like this: Because of the earth's rotation, winds in the Northern Hemisphere are deflected to the right; and winds of the Southern Hemisphere are deflected to the left.)

Since this Coriolis force affects every wind and every weather situation, pilots should understand it.

Figure 6–3 represents the earth rotating from west to east. The equator, four parallels of latitude, and a meridian of longitude are shown. The meridian is shown in two successive positions, the solid line indicating the meridian's position at a given instant, and the dotted line the same meridian's position at a later time. Centers of high and low pressure are indicated by H_1, H_2, H_3, and L_1, L_2, respectively.

On account of the difference in their respective distances from the earth's axis, the linear velocity of H_1 at the equator will be about 1000 mph, while that of L_1 and L_2 at 30° north and south latitude will be about 866 mph, and that of H_2 and H_3 at latitude 60° will be only about 500 mph.

Suppose that an air particle, under the effect of the pressure gradient, starts from H_1 toward L_1. Considering the particle to be in the free air—that is, above the effect of surface friction—the particle will continue to move toward the east with the same speed as its point of

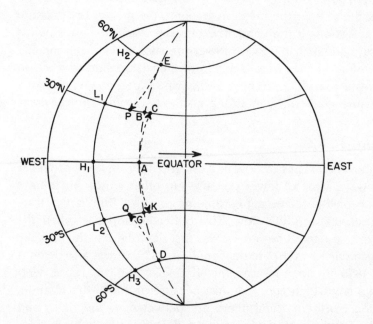

Figure 6–3 Effect of the Earth's Rotation: the Coriolis Force.

origin, point H_1, regardless of the velocity imparted to it in a northerly direction by the pressure gradient. While the particle is proceeding from the equator up to the latitude of L_1, the earth will rotate so as to bring the meridian to the position shown by the dotted line. Point H_1 will have moved to A, L_1 to B, and H_2 to E. The air particle, however, in retaining its original eastward component of velocity, will have moved to the eastward a *distance* equal to the travel of point H_1—that is, H_1–A—and consequently will arrive at a point C, to the eastward of B, such that L_1–C equals H_1–A. Thus we see that the path of the particle *relative to the rotating earth* is indicated by the broken arrow A–C. The particle, instead of following directly along the meridian A–B, has been deflected to the *right*.

Similarly, a particle starting from H_2 will, on account of its *smaller* eastward component, tend to lag behind the motion of the meridian as the particle proceeds southward. Thus, when point L_1 arrives at B, the air particle will arrive at some point, P, to the *westward* of B, so that $L_{-1}P$ equals H_2–E. Again we see that the particle has been deflected to the right of the direction of the decreasing barometric gradient.

In the Southern Hemisphere, the reverse is true, and air flowing from a higher to a lower pressure area will be deflected to the *left* under the influence of the Coriolis force.

Since the deflecting force tends to keep the particle from moving directly toward the area of low pressure, we may accept without further proof the fact that one component of the deflecting force always acts directly *away* from the low pressure, along the line of the *increasing* barometric gradient. Since the pressure force acts along the direction of the *decreasing* gradients, and this is also always at right angles to the isobars, the pressure force and the outward component of the deflecting force act in exactly opposite directions. The air particle will therefore continue to be deflected until the latter component and the pressure force become equal and opposite. (See Figure 6–4.) Between such balanced forces the air particle can move only at right angles to both of them—that is, *along the isobar at that point*—and this it does under the force supplied by the remaining components of the deflecting force. So long as these conditions continue to exist, the air will continue to flow parallel to the isobars much as a river flows between its banks. If the balance of forces becomes upset in any way, the particles then move toward, or away from, the low

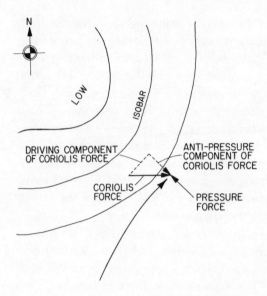

Figure 6–4 "Balanced" Condition of the Coriolis Force and the Pressure Force. In this condition the flow of air is parallel to the isobars.

pressure. Thus, we see that the circulation of the air about an area of low pressure (Figure 6–5) has a counterclockwise rotation in the Northern Hemisphere, and a clockwise rotation in the Southern Hemisphere. The rotation about an area of high pressure is the reverse in each case.

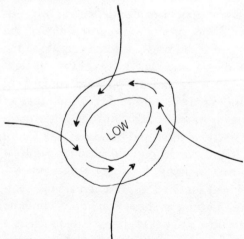

Figure 6–5 Circulation about an Area of Low Pressure Due to the Coriolis Force (Northern Hemisphere.)

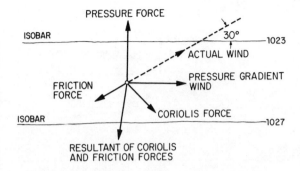

Figure 6–6   Resultant of Coriolis and Friction Forces.

FRICTION

The third effect of importance in determining wind strength and direction is *friction*. Friction retards air flow depending on the nature of the surface over which the air is moving. Since half of the earth's atmosphere is below 18,000 ft, it follows that the great mountain ranges which tower to this altitude, and far above, produce great changes in air circulation and offer great resistance to it. Thus friction is least over water and greatest over mountainous terrain. Also, friction takes place largely near the earth's surface, not at altitude. Over average terrain, the effect of friction extends only to about 2000 or 3000 ft. Above 3000 ft the actual wind and the pressure gradient wind are normally the same both in speed and direction. However, at the surface over land, the wind will flow across the isobars *toward* lower pressure at about a 30° angle and a speed of about one-half the pressure gradient wind speed as shown in Figure 6–6.

TYPES OF WIND

Pilots should know and understand these two types of winds:

Gradient wind—a wind flowing horizontally along and parallel to
        the isobars.
Geostrophic wind—a horizontal wind whose gradient force is balanced by the Coriolis force.
A gradient wind is illustrated in Figures 6–7 and 6–8.

For practical purposes, when the air flow has little curvature, the *gradient wind* and the *geostrophic wind* are considered equivalent and can be assumed to exist near and above 2000 or 3000 ft.

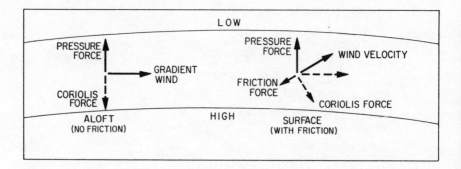

Figure 6–7   Comparison of Gradient and Surface Winds.

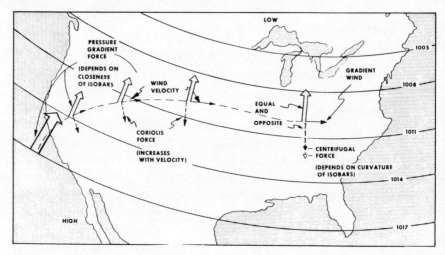

Figure 6–8   Gradient Wind.

Geostrophic Wind

The velocity attained by the wind when blowing under the conditions of complete balance of forces is called the geostrophic wind. Its velocity depends upon several factors: the pressure gradient force, the density of the atmosphere, the angular velocity of the earth's rotation, and the latitude. Since the atmospheric density is almost constant in a horizontal plane, the velocity of the geostrophic wind is directly proportional to the pressure gradient. Thus the geostrophic wind will be strongest with a steep pressure gradient, which, on a map representing distribution of pressure, may be readily recognized by the fact that the isobars are crowded closely together. Again the flow of the air between the isobars may be compared to the flow of a river between its banks: The narrower the river, the swifter its current.

It should be borne in mind that the geostrophic wind blows, or is approached, only in the "free atmosphere" well above the surface, from 2000 to 3000 ft (and more) of altitude. The retarding effect of ground friction always makes the surface wind less than the geostrophic wind.

CIRCULATION OF THE ATMOSPHERE

Having discussed the causes of wind and the forces which produce it, let us now examine the circulation of the atmosphere as a whole.

Circulation is the movement of air over the surface of the earth. This movement is continuous throughout the entire atmosphere, but pilots are primarily concerned with the movement of air in the troposphere. This circulation is caused by the unequal distribution of the sun's energy on the earth.

As might be expected from its close relation to the temperature, the pressure variations in the atmosphere closely follow the sun's annual motion, as explained in Chapter 3, page 23. In the Northern Hemisphere in summertime, more heat is absorbed from the sun than in the Southern Hemisphere. Moreover, in either hemisphere, the pressure over the land during the winter season is decidedly above the annual average, and during the summer season decidedly below it. The extreme variations occur on the continent of Asia, where the mean monthly pressure drops from 1033 mb in January to 999 mb in July. Over the North Atlantic and North Pacific oceans, on the other hand, conditions are reversed, the summer pressures being somewhat the higher. Thus, in January the Icelandic and the Aleutian lows increase in depth to 999 mb, while in July these minima fill up and are nearly obliterated. This fact has much to do with the strength and frequency of the winter gales in high northern latitudes, and the absence of gales during the summer.

Three-Cell Theory of Circulation

Many meteorologists accept the so-called three-cell theory of circulation. In this theory, the earth is divided into six latitude belts: three in the Northern Hemisphere and three in the Southern Hemisphere. The dividing lines between these six belts are:

(1) the equator

(2) 30° latitude north and south

(3) 60° latitude north and south

Consequently, there is one belt in each hemisphere between the equator and 30° latitude, one between 30° latitude and 60° latitude, and one between 60° latitude and the pole. (See Figure 6–9.)

Equator to 30° Latitude Cell　The air at the equator is heated and rises, thus creating a low-pressure area at the surface. When this warm air reaches the extremity of the atmosphere, it tends to flow toward the poles. About the time that this air has migrated to 30° latitude, the Coriolis force has deflected it so much that it is moving eastward instead of north. This results in a piling up of air near 30° latitude, and consequently, there is a descending current to the surface. When this descending air reaches the surface, part of it flows toward the equator to replace that which has gone and part of it flows northward toward the pole in the Northern Hemisphere. (The converse is true in the Southern Hemisphere.) As a result of this piling up of air in the

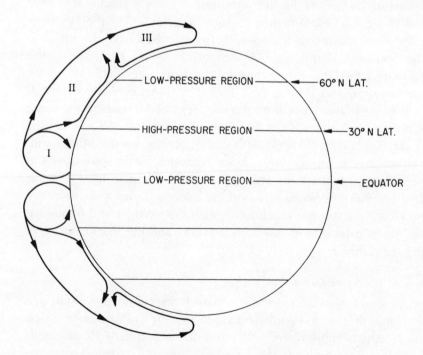

Figure 6–9　Three-cell Circulation Pattern.

area of 30° latitude, a belt of high pressure is formed in this area, which is one of the three cells of circulation.

30° Latitude to 60° Latitude Cell   The second cell of circulation is located between 30° latitude and 60° latitude. The circulation pattern in this cell is much more involved than the pattern in the first cell, since the flow of air is complicated by traveling cyclones and anticyclones. Although the general flow of air in the second cell is toward the poles, it is deflected to the east by the rotation of the earth (Coriolis force). This cell is known as the belt of prevailing westerlies.

60° Latitude to Pole Cell   The third cell of circulation lies between 60° latitude and the pole. The circulation in this cell begins with a flow of air at a high altitude toward the poles. Air which is forced to rise by the circulation in the second cell and by the heating in this area adds to the primary flow of air from the first cell. This flow of air continues to the poles, where it is cooled, and descends, forming a more or less static high-pressure area in the polar regions. When the air reaches the surface of the earth, it tends to flow back toward the equator near the surface of the earth. The circulation in this cell is completed by the southerly flow of air toward the equator. This air converges with the poleward flow from the second cell and is deflected upward. This convergence causes a semipermanent low-pressure area (at approximately 60° latitude), and the discontinuity in temperature and density of the two bodies of air causes a front (called the *polar front*) to form in this area. (See Figure 6–10.)

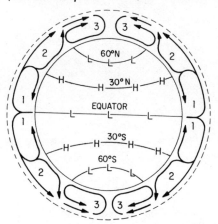

Figure 6–10 World Circulation Pattern.

THE WORLD'S WIND BELTS

As a result of the distribution of pressure just described, there is in either hemisphere a continual motion of the surface air away from the subtropical anticyclones—on one side toward the equatorial belt, on the other side toward the pole. (See Figure 6–11.) The first constitutes, in each case, the "trade winds," and the second, the prevailing winds of higher latitudes. As previously stated, the direction of this motion upon a stationary earth would be directly from the region of high toward the region of low pressure, the moving air steadily following the pressure gradient, increasing in force to a gale where the gradient is steep, decreasing to a light breeze where the gradient is gentle, and sinking to calm where it is absent. The earth, however, is in rapid rotation, and this rotation gives rise to the Coriolis force which diverts winds to the *right* in the Northern Hemisphere, and to the *left* in the Southern Hemisphere. The air set in motion by the difference of pressure is thus constantly turned aside from its natural course down the barometric gradient or slope, and the direction of the wind at any point is deflected by a certain amount, crossing the latter at an angle which varies between 45° and 90°, the wind in the latter case blowing parallel to the isobars. As a consequence of this deflection, the winds which one would naturally expect to find northerly on the equatorial side of the belt of high pressure at 30° latitude in the Northern Hemisphere become northeasterly to form the north-

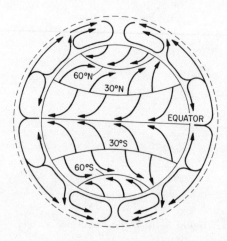

Figure 6–11 Direction of Prevailing Winds on the Earth's Surface.

east trades, while the winds on the poleward side of this belt, instead of being southerly, become southwesterly, forming the "prevailing westerlies" of the northern latitudes. So, too, for the Southern Hemisphere, the southerly winds of the equatorial slope become southeasterly—the southeast trades—and the northerly winds of the (south) polar slope of the belt become northwesterly, forming the "prevailing westerlies" of the southern latitudes.

The Trades

The trade winds blow from the subtropical belts of high pressure toward the equatorial belt of lower pressure—in the Northern Hemisphere from the northeast, in the Southern Hemisphere from the southeast. Over the eastern half of each of the great oceans they originate much farther away from the equator and their original direction is more nearly from the poles than in midocean, where their direction is almost easterly. They are commonly considered to be the most constant of winds, but although they may blow for days or even weeks with slight variation in direction or strength, their uniformity should not be exaggerated. At times these trade winds weaken or shift. There are regions where their steady course is deformed, notably among the island groups of the South Pacific, where, during January and February, the trades are practically nonexistent. They attain their highest development in the South Atlantic and the South Indian Oceans, and are everywhere fresher during the winter season than during the summer. They are rarely disturbed by cyclonic storms; the occurrence of cyclonic storms within the limits of the trade-wind region is confined, in time, to late summer and autumn months of the respective hemispheres, and, in location, to the western portion of the several oceans. The South Atlantic Ocean alone, however, enjoys complete immunity from tropical cyclonic storms.

The Doldrums

The doldrums is a belt of low pressure along the equator where light winds vary with calms. It is located between the subtropical anticyclones of the Northern and Southern Hemispheres. The name "doldrums" goes back to the days of sailing ships and probably is a corruption of the word "dull." In this belt, sailing vessels often became becalmed for days. Throughout this barometric belt or trough the pressure, save for the slight diurnal oscillation, is practically uniform, and decided barometric gradients do not exist. Here, accordingly, the winds sink to stagnation or, at most, rise only to the strength of fitful breezes, coming first from one direction, then from another, with cloudy, rainy sky and frequent thunderstorms. The region throughout which these conditions prevail consists of a wedge-shaped area, the base of the wedge resting, for the Atlantic Ocean, on the west coast of Africa, and, for the Pacific Ocean, on the west coast of America, the axis of the wedge extending westward. Throughout February and March it is found immediately north of the equator and is of inappreciable width; vessels following the sailing routes frequently pass from trade to trade without interruption in both the Atlantic and Pacific Oceans. In July and August it has migrated to the northward; the axis extends east and west along the parallel of about 7° north latitude, and the belt itself covers several degrees of latitude, even at its narrowest point. At this season of the year, also, the southeast trades blow with diminished freshness across the equator and well into the Northern Hemisphere. They are diverted, however, by the effect of the earth's rotation into southerly and southwesterly winds, the so-called southwest monsoon of the African and Central American coasts.

The Horse Latitudes

On the outer margin of the trades, corresponding vaguely with the summit of the subtropical ridge of high pressure in either hemisphere, is a second region throughout which the barometric gradients are faint and undecided, and the prevailing winds correspondingly light and variable. These are the so-called "horse latitudes" or calms of Cancer and Capricorn. It is generally supposed that these latitudes got their unusual name in colonial times when sailing ships carrying horses to the West Indies became becalmed for so long that the horses' food and water were exhausted and the animals had to be thrown overboard. Unlike the doldrums, however, the weather here is clear and fresh; the periods of stagnation are intermittent rather than continuous and show none of the persistency which is so characteristic of the equatorial region. The explanation of this difference will become obvious when we realize the nature of the daily barometric changes of pressure in the respective regions, for while the doldrums are marked by the uniformity of the torrid zone with but slight diurnal variation, the horse latitudes share to a limited extent in the wide and rapid variations of the temperate zone.

The Prevailing Westerlies

On the exterior or polar side of the subtropical anticyclones the pressure again diminishes; the barometric gradients are here directly toward the pole. The currents of air set in motion along these gradients, diverted to the right and left, respectively, of their natural course by the earth's rotation, appear in the Northern Hemisphere as southwesterly winds and in the Southern Hemisphere as northwesterly. These are called the "prevailing westerlies" of the temperate zones because they blow from one direction more than from any other.

Only in the Southern Hemisphere do these winds exhibit anything approaching the persistency of the trades. Their course in the Northern Hemisphere is subject to frequent local interruptions by periods of winds with an easterly component. Tabulated results show that throughout the portion of the North Atlantic included between latitude 40° and 50° north and longitude 10° to 50° west, winds from the western semicircle comprise about 74% of the whole number of observations. The relative frequency is somewhat higher in winter, somewhat lower in summer. The average force, on the other hand, decreases from force 6 to force 4, Beaufort scale (see page 77), with the change of season. Over the sea in the Southern Hemisphere such variations are not apparent; here the westerlies blow through the entire year with a steadiness little less than that of the trades themselves and with a force which, though fitful, is very much greater; because of their boisterous nature the name the "roaring forties" has been given to the latitude in which they are most frequently observed.

The explanation of this striking difference in the extratropical winds of the two halves of the globe is found in the distribution of atmospheric pressure, and in the variations of this in different parts of the world. In the nearly landless Southern Hemisphere the atmospheric pressure after crossing the parallel of 30° south diminishes almost uniformly toward the pole, and is rarely disturbed by those large and irregular fluctuations which form so important a factor in the daily weather of the northern hemisphere. Here, accordingly, there is a system of polar gradients quite comparable in stability with the equatorial gradients which give rise to the trades, and the poleward movement of the air in obedience to these gradients, constantly diverted to the *left* by the effect of the earth's rotation (Coriolis force), results in the steady westerly winds of the South Temperate Zone.

GENERAL CIRCULATION PATTERN

The earth's general circulation pattern is shown in Figures 6–12 and 6–13. In the Northern Hemisphere, near 30° north latitude, there are two high-pressure cells, one over the Pacific and one over the Atlantic —uniform in substance. This follows along with the general circulation theory of the subtropical high-pressure belt; however, over land at this latitude (due to low specific heat of land and its more immediate response to insolation) the theory breaks down, and changeable pressure patterns can be expected here.

Near 60° latitude, two high-pressure cells are found over continental areas—one over Canada, the other over Siberia. In summer, due to long hours of daylight in northern regions (and also due to the fact that the doldrum area will have moved north of the equator), the high-pressure belt retreats to a more localized position near the pole.

In the Aleutian and Icelandic areas, the Aleutian low and Icelandic low also bear out the polar front or low-pressure theory in the latitude

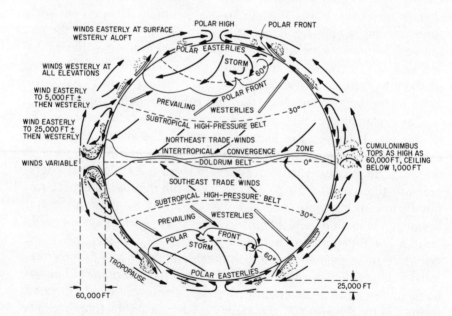

Figure 6–12 General Atmospheric Circulation

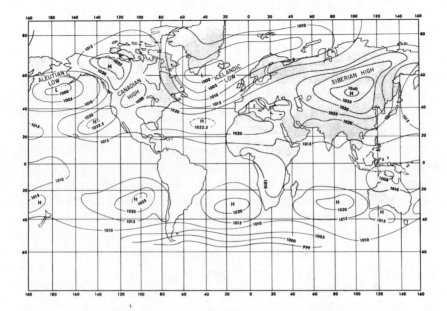

Figure 6–13 Prevailing World Pressure Systems.

of 60° north. (See Figure 6–13.) This is the area where the warmer air of the prevailing westerlies (northern portions of the subtropical highs) is forced to ascend over the more cold and dense air of the continental highs.

To draw an analogy about these lows, they act like planetary gears between and among the four high-pressure cells; the highs have winds rotating clockwise, and to prevent clashing of the gears, there must be some lows in the system rotating counterclockwise.

Every pilot should be familiar with the location of these six semi-permanent pressure systems: the Siberian, Pacific, Canadian, and Bermuda highs and the Icelandic and Aleutian lows, because they play an important part in the origination and movement of air masses and frontal systems, which will be studied in succeeding chapters.

LOCAL WINDS

Monsoons

This is the name given to local seasonal winds. (The name comes from the Arabic word *mausim* meaning season.) While most people associate monsoons with Arabia and India, monsoons also occur in many other places—Spain, Australia, and Chile, to mention a few.

Air over land is warmer in summer and colder in winter than that over adjacent oceans. During the summer the continents become the seat of areas of relatively low pressure and, during the winter, of relatively high. Pressure gradients directed outward from the continents during the winter, inward during the summer, are thus established between the land and the sea; these exercise great influence over the winds prevailing in the region adjacent to the coast. Thus, off the Atlantic seaboard of the United States southwesterly winds are most frequent in summer, northwesterly winds in winter; but on the Pacific coast, the wind changes from northwest in summer to southwest with the advance of winter.

The most striking monsoon winds of this class are those of the Southeast Asia and the Indian Ocean. In January abnormally low temperatures and high pressure prevail over the Asiatic plateau, high temperatures and low pressure over Australia and the nearby portion of the Indian Ocean. As a result of the barometric gradients thus established, the southern and eastern coasts of the vast Asiatic continent and the adjacent seas are swept by an outflowing current of air which, diverted to the right of the gradient by the earth's rotation, appears as a northeast wind, covering the China Sea and the northern Indian Ocean. As these winds enter the Southern Hemisphere, however, the same force which hitherto deflected them to the right of the gradient now serves to deflect them to the left. Here, consequently, the monsoon appears as a northwest wind covering the Indian Ocean as far south as 10° south latitude, the Arafura Sea, and the northern coast of Australia.

In July these conditions are exactly reversed. Asia is now the seat of high temperature and correspondingly low pressure, and Australia of low temperature and high pressure, although the departure from the annual average is by no means so pronounced in Australia as in Asia. The pressure gradients thus lead across the equator and are directed toward the interior of the greater continent, giving rise to a system of winds whose direction is southeast in the Southern Hemisphere and southwest in the Northern Hemisphere.

The northeast (winter) monsoon blows in the China Sea from October to April, the southwest (summer) monsoon from May to September. The former is marked by all the steadiness of the trades, often attaining the force of a moderate gale; the latter appears as a light breeze, unsteady in direction, and often sinking to a calm. Its prevalence is frequently interrupted by tropical cyclonic storms, locally known as *typhoons,* although these may occur well into the winter monsoon season.

One important effect of the seasonal variation of temperature and pressure over the land remains to be described. If there were no land areas to break the even water surface of the globe, the trades and westerlies of the general terrestrial circulation would be developed in the fullest simplicity, with linear divisions along latitude circles between the several members—a condition nearly approached in the land-barren Southern Hemisphere during the entire year, and in the Northern Hemisphere during the winter season. In the summer, however, the subtropical belt of high pressure is broken where it crosses the warm land, and the air pushed off from the continents accumulates over the adjacent oceans, particularly in the northern or land hemisphere. This tends to create over each of the oceans a circular or elliptical area of high pressure, from the center of which the pressure gradients radiate in all directions, giving rise to an outflowing system of winds, which by the effect of the earth's rotation is converted into an outflowing spiral eddy or anticyclonic whirl. The sharp lines of demarcation which would otherwise exist between the several members of the general circulation are thus submerged and obliterated. The southwesterly winds of the middle northern latitudes become successively northwesterly, northerly, and northeasterly, as we approach the equator in following the eastern edge of the high-pressure area; and the northeast trade becomes successively southeasterly, southerly, and southwesterly as we continue on around the western edge of the area, receding from the equator. The exact reverse will, of course, be true for the other hemisphere.

Land and Sea Breezes

These phenomena are observed along the coasts of large bodies of water and are more pronounced in summer than in winter. By the effects of solar radiation, the land heats more rapidly during the day than the water, and cools more rapidly during the night. The rising convective currents over the land cause the cooler air over the water to flow in to take its place, producing a wind that blows from the sea, called the *sea breeze.* At night the reverse condition develops. (See

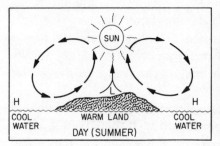 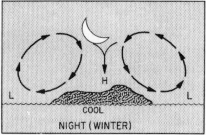

Figure 6–14   Land-sea Effect.

Figure 6–14.) The land cools much more rapidly and soon is at considerably lower temperature than the water. The air over the land, cooled by contact, becomes denser and flows out over the water, forcing the warmer and lighter air upward, thus creating a wind that blows from the land, called the *land breeze*. These breezes are confined to a rather narrow area along the coast, seldom extending more than about 15 miles inland, although the breezes may be felt considerably farther to seaward, particularly during the daytime. The sea breeze, depending upon the degree of heating of the land surface, may develop into a fresh or strong breeze; the land breeze at night is usually gentle.

In considering, observing, and analyzing the weather it must be borne in mind that these conditions are superimposed upon the general conditions and one should not, therefore, be misled by them. They will be most pronounced, of course, in the tropics, where land and sea breezes often blow with great persistence; but, in latitudes farther from the equator, stronger winds of a more general character at times may completely prevent the occurrence of these phenomena.

Valley and Mountain Breezes

On warm days, winds tend to blow up slopes during the day and down slopes during the night. This is because mountain slopes are warmer than the free atmosphere at the same level during the day and colder during the night. Since cold air tends to sink and warm air tends to rise, a system of winds develops which flows up the mountain during the day and down during the night. The daytime movement is called a *valley breeze;* the nighttime motion is called a *mountain breeze.*

Fall Winds

In some places in the world, where a very cold inland plateau is adjacent to a coastal region, the force of gravity may cause heavy, dense, cold air to drain off or flow down to low levels. These fall winds—also called *katabatic* winds—can last for several days and can be quite strong—more than 100 knots. The mistral of the northwest coast of the Mediterranean and the bora of the Adriatic coast are examples of such fall winds. If the down-slope wind is warm, relative to the plain or valley, it is called a *foehn* wind. The chinook of the Rockies and the Santa Ana of California are examples.

THE BEAUFORT SCALE

In 1805, a British admiral, Sir Francis Beaufort, worked out a method for classifying the apparent strength of winds. Admiral Beaufort based his method on thirteen groups of wind according to velocity. The classification numbers, ranging from 0 to 12, are called *Beaufort numbers*. Although nautical miles per hour (knots) are used extensively today, adaptations and variations of the Beaufort scale are still used by certain navigators and some meteorological organizations preparing forecasts. (See Table 6–1.)

Table 6–1

BEAUFORT NUMBER	DESCRIPTIVE TERM	VELOCITY EQUIVALENT AT A STANDARD HEIGHT OF 10 METERS ABOVE OPEN FLAT GROUND			
		Mean Velocity			
		In Knots	*Meters/Sec*	*km/hr*	*mph*
0	Calm	< 1	0–0.2	< 1	< 1
1	Light air	1–3	0.3–1.5	1–5	1–3
2	Light breeze	4–6	1.6–3.3	6–11	4–7
3	Gentle breeze	7–10	3.4–5.4	12–19	8–12
4	Moderate breeze	11–16	5.5–7.9	20–28	13–18
5	Fresh breeze	17–21	8.0–10.7	29–38	19–24
6	Strong breeze	22–27	10.8–13.8	39–49	25–31
7	Near gale	28–33	13.9–17.1	50–61	32–38
8	Gale	34–40	17.2–20.7	62–74	39–46
9	Strong gale	41–47	20.8–24.4	75–88	47–54
10	Storm	48–55	24.5–28.4	89–102	55–63
11	Violent storm	56–63	28.5–32.6	103–117	64–72
12	Hurricane	64 and over	32.7 and over	118 and over	73 and over

7 CLOUDS

DEFINITION OF CLOUDS

Clouds are countless millions of minute water droplets, or ice particles, suspended in the atmosphere above the earth's surface. These microscopic specks of ice or water have a central core or particle of dust, ash, salt, or smoke (called *nuclei*). The diameter of these nuclei varies from .0001 to .004 in., and they are easily supported and carried by air movements as slow as one-tenth of a mile per hour. The World Meteorological Organization definition of cloud is "a visible aggregate of minute particles of water or ice, or of both, in the free air."

Clouds are the signposts of the sky. They are visible evidence of what the atmosphere is doing in terms of motion and water content. Much of what we discuss as "weather" is actually a discussion of cloud formations which accompany various atmospheric movements.

Clouds differ from fog only in that the base of fog must be within 50 ft of the ground. If you look at a mountain, the top of which is obscured, you would of course say that the mountaintop is lost in a cloud. But to a person who is up in the mountain, it would be fog.

Clouds can be composed of water vapor and ice crystals, or solely of ice crystals.

For the pilot, the knowledge of clouds, their types, classifications and names, how they are formed, and their significance is particularly important in the planning and conduct of a safe flight. Such knowledge is also essential to gain full use of Surface Weather Maps, Prog Charts, and Weather Depiction Charts. The aviator who can analyze the clouds he encounters in flight will be able to make reasonable predictions about flight conditions further ahead. Also, by observing cloud motions a pilot can obtain information about the velocity and direction of winds aloft. Moreover, if a pilot can accurately report to the meteorologist about the clouds he sees during flight, or after landing, valuable information will be provided to other pilots.

HOW CLOUDS ARE FORMED

When air can no longer hold all of its water vapor, a cloud is formed by the condensation of a part of the water vapor into visible droplets, snow, or ice crystals. Three conditions are essential to cloud formation:

(1) Sufficient water vapor must be present.
(2) There must be a cooling process.
(3) There must be nuclei, or core particles, around which the water droplets can form.

There are several processes by which the air is cooled and clouds formed: (1) convective cooling by expansion, (2) mechanical cooling by expansion, and (3) radiational cooling. Any of the three methods may be working in conjunction with another method, thus making it even more effective.

Convective Cooling

The ascent of a limited mass of air through the atmosphere due to surface heating is called thermal convection. If a sample of air is heated, it rises (being less dense than the surrounding air) and decreases in temperature until the surrounding air has a temperature equal to, or higher than, the sample of air. Then convection ceases. Cumuliform clouds are formed by this means with their bases at the altitude of saturation and their tops at the point where the temperature of the surrounding air is the same as, or greater than, the temperature of the sample of air.

Mechanical Cooling

Mechanical cooling is of two types—orographic and frontal.

Orographic If air is comparatively moist and is lifted by the wind over mountains or hills, clouds may often be formed. The type of cloud depends upon the lapse rate (the rate of decrease in temperature with increase in height unless otherwise specified) of the surrounding air. If the lapse rate is weak (that is, a small rate of cooling with an increase in altitude), the clouds formed are of the stratiform type. If the lapse rate of the surrounding air is steep (that is, a large rate of cooling with increasing altitude), the clouds formed are of the cumuliform type.

Frontal In the discussion of clouds being formed by frontal slopes, we have two huge masses of air, each having its own characteristic properties. Due to the difference of density, the two air masses do not readily mix. Warm air is less dense than cold air. Therefore, the warm air moves up over the cold air. The cold air acts as an inclined plane, and the boundary between the two masses of air is called the frontal surface.

If warm air flows slowly up over a cold mass of air, the air is cooled adiabatically (no transfer of heat or mass) to its dew point and stratified clouds are formed. On the other hand, if cold air pushes under warm air, the warm air is forced aloft rapidly, and upon condensation, cumuliform clouds result.

Radiational Cooling

At night the earth reradiates long-wave radiation, thereby cooling rapidly. The air in contact with the surface is not heated by the outgoing radiation but, instead, is cooled by contact with the cold surface. This contact cooling lowers the temperature of the air near the surface, causing a surface inversion. If the temperature of the air is cooled to its dew point, fog and/or low clouds form. Clouds formed in this manner dissipate during the day due to surface heating.

Clouds can also form in air which is already at or close to its dew point by rain falling through it from higher clouds.

The third requirement for cloud formation is the presence of a core, or nucleus, around which condensation can begin. Such cores are present in the atmosphere from: (1) salt particles from sea spray which float in the atmosphere in great quantities; (2) smoke particles from the incomplete combustion of industrial wastes, automobile exhausts, dust, forest fires, and volcanoes.

HOW CLOUDS ARE CLASSIFIED

There are a number of methods of cloud classification. Clouds may be classified according to appearance, how they are formed, or the height of their bases.

But it was not until 1803 that an Englishman named Luke Howard made the first classification of cloud forms. Howard gave the three main classes Latin names—*cirrus* meaning curly, *stratus* meaning spread out, and *cumulus* meaning heaped up.*

From this beginning, an international committee met in 1896 and decided upon ten classes of clouds (called genera):

GENERA	ABBREV.	AVERAGE HEIGHT IN MID-LATITUDE REGIONS
Cirrus	(Ci)	
Cirrocumulus	(Cc)	20,000 ft and higher
Cirrostratus	(Cs)	
Altocumulus[a]	(Ac)	
Altostratus	(As)	6500 ft to 20,000 ft
Stratocumulus	(Sc)	
Stratus	(St)	Surface to 6500 ft
nimbostratus[b]	(Ns)	
Cumulus	(Cu)	
cumulonimbus[c]	(Cb)	1600 ft to cirrus level

[a] Alto—from the Latin *altum*, or height.
[b] Nimbo—from the Latin *nimbus*, which means rainy cloud.
[c] Because of their turbulence, hail, heavy rain and thunder, the cumulonimbus (Cb) is referred to by pilots as "Cb," pronounced "sea-bee."

---

* The actual etymology of these weather terms from Latin is as follows:
    cirrus—from the Latin *cirrus*, which means a lock of hair, a tuft of horsehair, or a bird's tuft.
    stratus—from the Latin verb, *sternere*, which means to extend, to flatten out, to spread out, or to cover with a layer. Stratus is the past participle of *sternere*.
    cumulus—from the Latin word *cumulus* which means an accumulation, pile or heap. In speaking, pilots often refer to cumulus clouds simply as "cue."

The Ten Genera of Clouds.

In 1957, the World Meteorological Organization published the *International Cloud Atlas,* in two volumes with colored pictures, which has become the "bible" on clouds. In this, there are three *stages* (levels or *etages*), ten *genera,* fourteen *species,* nine *varieties,* and nine supplementary features which classify clouds. For the pilot, however, knowledge of the ten classes above is sufficient. The stages or etages are as follows:

ETAGE	POLAR REGIONS	TEMPERATE REGIONS	TROPICAL REGIONS
High	3–8 km	5–13 km	6–18 km
	(10,000–25,000 ft)	(16,500–45,000 ft)	(20,000–60,000 ft)
Middle	2–4 km	2–7 km	2–8 km
	(6500–13,000 ft)	(6000–23,000 ft)	(6500–25,000 ft)
Low	From the earth's	From the earth's	From the earth's
	surface to 2 km	surface to 2 km	surface to 2 km
	(6500 ft)	(6500 ft)	(6500 ft)

TYPES AND DESCRIPTIONS OF CLOUDS

High Clouds

Cirrus (Ci) (Figures 7–1, 7–2, and 7–3)   Cirrus is a detached cloud of delicate and fibrous texture, usually without shading, generally white in color, often of a silky appearance, always composed of ice crystals and sometimes called mare's tails.

Figure 7–1 Cirrus Uncinus. These clouds are in parallel trails and small patches. (Official photograph, Environmental Science Services Administration.)

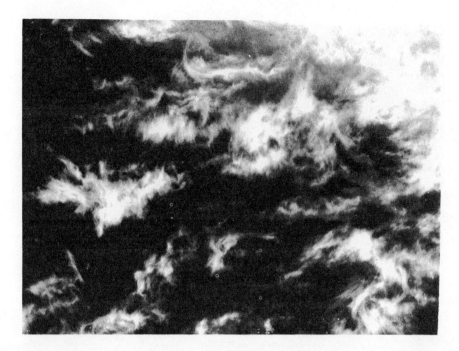

Figure 7–2   Cirrus Fibratus. This is cirrus with an irregular arrangement of filaments. (Official photograph, Environmental Science Services Administration.)

This is the highest and thinnest of all clouds, occurring generally at altitudes above 20,000 ft where the extremely low temperature (−30°F. to −40°F.) causes the formation of ice crystals instead of water droplets. Because of its extreme thinness, and because of the various processes operating to produce it, this cloud appears in a greater variety of forms than any other. The amount of water vapor at this elevation varies in the same way as at all other altitudes. When the amount of water vapor is small, the cirri appear as delicate threads. As the quantity of moisture increases, they appear as tufts and tangled masses, or even as sheets and streams.

When detached, light and silky in appearance, without connection with cirrostratus or altostratus, cirrus is usually considered a sign of fair and settled weather. This type is illustrated in Figure 7–1.

Certain other cirrus types are usually associated with lows. Cirri arranged in long parallel bands, which because of perspective seem to radiate from a definite point over the horizon, can be the forerunner of a storm. (An ancient rhyme of the sea goes, "Mackerel sky and mare's tails, make tall ships carry low sails.") Cirri composed of

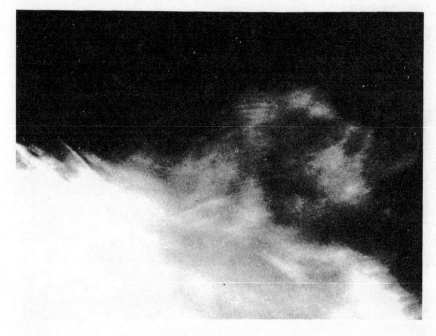

Figure 7–3 Cirrus above 45°. (Official photograph, Environmental Science Services Administration.)

drawn-out filaments with tufts or tangled masses at their ends warn of an approaching low. These so-called mare's tails can be followed very shortly by cirrostrati and then by the lowering altostrati, indicating definitely the approach of a storm within 24 to 48 hours.

Before sunrise and after sunset, cirrus clouds may still be colored bright yellow or red. Being high-altitude clouds, they light up before lower clouds and fade out much later.

Cirrostratus (Cs) (Figure 7–4)   Cirrostratus appears in patches as a thin whitish sheet covering the sky with a whitish veil. Such a sheet often produces halos around the sun and moon, as shown in Figure 7–4. This is because the ice crystals refract the light which passes through them in such a way as to produce the characteristic ring. Cirrostratus is similar to the cirrus but is lower, thicker, and forms a more uniform sheet. It is composed of ice crystals, and its average altitude is about 28,000 ft.

Like cirrus, cirrostratus sometimes forms as a result of the evapora-

Figure 7–4 Cirrostratus Covering the Whole Sky, Giving the "Halo" Effect. (Official photograph, Environmental Science Services Administration.)

Figure 7–5 Cirrocumulus. (Official photograph, Environmental Science Services Administration.)

tion and subsequent recondensation at higher altitudes of cumulus and cumulonimbus. If cirrostratus clouds, which follow after cirrus, thicken and change to altostratus, such appearance is a good indication of rain, the probability being about 80% during the next 24 hours.

Cirrocumulus (Cc) (Figure 7–5)   Cirrocumulus occurs in small globular masses, or white flakes either without shadows or with very slight shadows, arranged in groups and often in lines. The average altitude is about 22,000 ft. It forms from a single or double undulation of cirrus or cirrostratus sheet which is to a certain degree unstable at its altitude. This instability, however, is frequently of very limited duration; cirrus and cirrostratus often are observed to pass quickly into cirrocumulus and then to reappear shortly after in their original form, with no evidence of cirrocumulus structure.

This type of cloud occurs frequently in its soft "flocculent" form,* as shown in Figure 7–5, and is usually an indication of fair weather. An indication of coming rain and wind, however, is found in that grayer and "harder" variety of cirrocumulus which resembles the scaly pattern on the backs of mackerel, and produces what is known as mackerel sky.

Medium or Middle Clouds

Altostratus (As) (Figure 7–6)   This is a striated, fibrous, or smooth veiled cloud, more or less gray or bluish in color like thick cirrostratus, but without a halo phenomenon. The sun or moon usually shows vaguely with a faint gleam, as through ground glass. This is popularly known as a watery sky.

Normally, this cloud is not itself a rain cloud, but is associated with the coming wet weather. It usually precedes a general rain by a few hours; in the northern part of the United States, this period is about six hours.

When cirrostratus follows cirrus within a few hours and is in turn followed by altostratus, the probability of rain within the succeeding six to twelve hours is about 90%, which is as great an accuracy as can be derived from a knowledge of the conditions prevailing over the country as gained from the weather map.

The approximate height of altostratus is about 13,000 ft, and its

---

* From Latin *floccus* which means tuft of wool, fluff, or nap of a cloth.

Figure 7-6   Altostratus. Known as a "watery sky," or a thick gray curtain, the alto-stratus may be followed in a few hours by continuous snow or rain. (Official photograph, Environmental Science Services Administration.)

thickness is considerably greater than one would ordinarily imagine, being from 1000 to 7000 ft thick at different times of the year.

Altocumulus (Ac) (Figures 7–7, 7–8, 7–9, and 7–10)   Altocumuli are rather large ball-like masses, white or grayish, partially shaded, arranged in groups or lines, and often so closely packed that their edges appear to touch. They are often mistaken for an unbroken layer of stratocumulus. The smaller elements of the regularly arranged units may be fairly small and thin, with or without shading appearing around the margin of a group whose center is generally larger and more compact. This is shown in Figures 7–7, 7–8, and 7–9.

Figure 7–10 shows the lenticular form of altocumulus, most frequently seen above mountains and high plateaus, where they are constantly blown by wind, and often molded into rolls.

Figure 7-7 Altocumulus Duplicatus. (Official photograph, Environmental Science Services Administration.)

Low Clouds

Stratocumulus (Sc) (Figures 7–11 and 7–12)   This cloud type develops into patches, or layers, of globular or roll form, the smallest of the regularly arranged elements being fairly large, soft, and gray with darker parts. It shows a low, continuous sheet also, which may be thick or thin with distinct irregularities of large size. The elements are arranged in groups, lines, or waves, aligned in one or two directions. Very often the rolls are so close together that the edges join. They may cover the whole sky with a wavy appearance.

Figure 7–11 is a picture of a stratocumulus from the surface, while Figure 7–12 is one from high altitude.

Stratocumulus is most frequent in the winter; in this season its altitude is much below the average—often 1000 ft or even less. This is a comparatively thin cloud (the average thickness is about 1400 ft) but when well developed it makes the day dull and dreary. It will sometimes "burn off" after the sun breaks through the thin portions and heats the earth below. At other times it may cover the sky for

Figure 7–8 Altocumulus Translucidus. (Official photograph, Environmental Science Services Administration.)

Figure 7–9 Altocumulus Translucidus Undulatus. These clouds are advancing over the sky in parallel bands. (Official photograph, Environmental Science Services Administration.)

Figure 7–10 Altocumulus Lenticularis. This is a cloud form seen mostly above mountains. (Official photograph, Environmental Science Service Administration.)

Figure 7–11 Stratocumulus Cumulogenitus. (Official photograph, Environmental Science Services Administration.)

Figure 7–12 Stratocumulus. (Official photograph, Environmental Science Services Administration.)

several days without breaking. Generally there are updrafts under the thicker portions which cause considerable bumpiness above and below the cloud, as well as within it. When it appears in rolls it is a good indication of wind, for the rolls really mark the crests of atmospheric waves. The rolls lie nearly at right angles to the wind producing them. If there is not much wind at the surface, it should be looked for just *above* the cloud layer.

Nimbostratus (Ns) (Figure 7–13) This is a low, amorphous layer, usually nearly uniform, and of a dark-gray color. It seems feebly illuminated, seemingly from the inside.

Nimbostratus usually evolves from a layer of altostratus which has thickened and developed downward, sometimes with a ragged appearance, and has sunk lower in the atmosphere. Beneath this layer there is generally a progressive development of very low, ragged clouds, separated at first, but later fused into an almost continuous layer. Broken, irregular fragments often drift below the layer, particularly as the storm is breaking up; these are called "scud."

Figure 7–13 Nimbostratus. (Official photograph, Environmental Science Services Administration.)

Nimbostratus is the ordinary rain cloud, though not the type which gives the heaviest rainfall. Rain or snow precipitated from nimbostratus is usually steady. The base of this cloud is seldom higher than 6500 ft above the earth; its altitude varies considerably, the most usual level being about 2000 ft. The duration of the cloud depends upon its size and development, and upon the rate of progression of the general storm conditions with which it is associated. It may last only a few hours, but, especially in winter, it may cover the sky for two or three days. A rising barometer and a change in wind are the usual indications of clearing.

Stratus (St) (Figures 7–14 and 7–15) This is a low, uniform layer of cloud resembling fog, but not resting on the ground. Figure 7–14 is a typical stratus cloud.

When a stratus cloud is thin, it gives the sky a characteristic hazy appearance. At other times, it may be a solid sheet of cloud; in fact, stratus may reach a thickness of several hundred feet. When broken,

Figure 7–14   Stratus. (Official photograph, Environmental Science Services Administration.)

Figure 7–15   Stratus Above an Island. (Official photograph, Environmental Science Services Administration.)

it is called fractostratus. It may occur with any other type of cloud. During or after precipitation, it frequently forms as scud beneath nimbostratus.

Stratus may be typical fog which has actually been lifted from the ground; it may also be formed by condensation between cool and warm currents. Figure 7–15 is such a cloud above an island.

Rain from stratus always takes the form of a light drizzle—that is, small droplets very close together. When there is no precipitation, stratus shows some contrasts of dark and light patches, the lighter parts being nearly transparent. Pure stratus (that is, stratus formed as such) usually does not "burn off" as easily as does typical land fog, and may persist for several days.

The aviator may gain valuable information while flying over stratus: Under favorable conditions, when the cloud sheet is not too thick, its upper surface indicates the contours of the ground beneath. Rivers and valleys appear as depressions, hills and mountains as elevations.

Figure 7–16 Cumulus. (Official photograph, Environmental Science Services Administration.)

This is a practical point worth remembering; many aviators, doubtful of their location, have been able to get their bearings in just this way.

A condition of *inversion* (that is, an *increase* of temperature with increase of altitude) often accompanies stratus; in such a case, the layer of cloud floats just below the stable warm layer of the inversion.

Clouds with Vertical Development

Cumulus (Cu) (Figures 7–16, 7–17, and 7–18)   Cumulus is a thick cloud with vertical development; the upper surface is dome-shaped, like cauliflower, exhibiting rounded protuberances, while the base is nearly horizontal.

When the cloud is opposite the sun, the surfaces toward the observer are brighter than the edges of the protuberances. When the light comes from the side, the cloud shows strong contrasts of light and shade. Seen against the sun the cloud is dark with bright edges. True cumulus is sharply defined above and below; its surface often appears hard and clear-cut. There is a ragged type of cumulus, however, called fractocumulus (Fc), which shows continual change (Figure 7–17). The base is usually gray. Even when large, cumulus produces only light precipitation, if at all.

Figure 7–17   Fractocumulus. The clouds are broken by turbulence and are "thinner" than ordinary cumulus. (Official photograph, Environmental Science Services Administration.)

Figure 7–18 Cumulus Humilis. (Official photograph, Environmental Science Services Administration.)

This type of cloud is always formed by the action of convective currents in the following manner: Air warmed by contact with the earth's surface under sunshine, becoming lighter, rises as a cork rises through water; thus an ascending current is created. This effect is most pronounced over the sunny slopes of mountains and unwooded areas, over plowed fields and sandy beaches. It is less pronounced over cool forest lands, fields covered with grassy crops, and over lakes. The current cools the warm air by expansion as it rises, finally reaching its dew point; a cloud begins to form at what is called the condensation level, which marks the flat base of the cloud. By its own inertia, sometimes with the aid of unstable conditions, the air rises beyond this level, and so the cloud grows vertically until halted by a layer of stable air.

Because they are the visible tops of invisible ascending currents, cumuli are of great interest to the aviator. Since rising warm air is replaced by descending colder air, downward currents also are to be expected. Cumuli indicate areas of turbulence, or "bumpiness," which the pilot should avoid. On the average, these turbulent areas extend between the altitudes of 4000 ft and 6000 ft, though they may reach as high as 10,000 ft.

Research shows that the rate of ascent of warm air currents in cumuli often reaches 10 to 12 ft/sec. In thunderstorms, the rate is much more rapid.

Although the most active ascending currents are practically certain to produce clouds, turbulence frequently does occur on cloudless days, when the air is dry. In such a case, the air is not cooled to its dew point until it reaches very high altitudes. We see now why those fine, warm, summer skies, which appear ideal for flying, turn out to be uncomfortably bumpy.

A type of cumulus which is flat and only slightly rounded on top, with little vertical development, and which shows only faint shadows, or none, on the under side, is called cumulus humilis.* This cloud is an indication of fine, fair weather. Cumulus humilis is illustrated in Figure 7–18.

Cumulus clouds which grow to great heights with typical "cauliflower" structure indicating great turbulence, internal motion, and congestion within, are called cumulus congestus. This type usually develops very rapidly into the cumulonimbus or thunderstorm cloud described in the next section.

Figure 7–19 Cumulonimbus. (Official photograph, Environmental Science Services Administration.)

* From the Latin *humilis,* which means near the ground, low, of small size.

Figure 7–20 A Cumulonimbus with Showers Beneath it. (Official photograph, Environmental Science Services Administration.)

Cumulonimbus (Cb) (Figures 7–19, 7–20, and 7–21) This is a heavy mass of cloud with great vertical development. The upper part rises like a tower or mountain, often spreading out in the form of an anvil, and has a fibrous texture.

Cumulonimbus generally produces showers of rain or snow, sometimes of hail, and often thunderstorms as well. A mass of cumulus, however heavy it may be and however great its vertical development, should not be classed as cumulonimbus unless at least part of its top is in the process of transformation into a cirrus mass. When a quick, heavy shower occurs, it may be assumed that the cloud is cumulonimbus even though lower clouds may hide the fibrous crown.

Figure 7–20 shows a fully developed cumulonimbus with rain already falling from it.

Figure 7–21 shows a typical cumulonimbus with its great anvil head spreading out over it. Note the ragged edge to the anvil and the wisps of cirrus about it. The anvil is composed chiefly of ice crystals which are extremely active nuclei for raindrops. Often such an anvil or tower may be seen in the sunlight from many miles away long before the base of the cloud becomes visible. This signpost in the sky marks an area of most turbulent air with probable hail and torrential rain of the "cloudburst" variety. *The sensible pilot never attempts to fly through or under such a storm but always goes around it.* Even in the

Figure 7–21 Cululonimbus Capillatus. (Official photograph, Environmental Science Services Administration.)

Figure 7–22 Cumulus Mammatus. (Official photograph, Environmental Science Services Administration.)

very tops of cumulonimbus the turbulence can be lethal. Since an air-craft's maximum attainable speed in level flight and its stalling speed are usually very close at these altitudes, the moderate turbulence can easily result in a stall. A stall from such a position forces the pilot to attempt almost impossible recovery while falling through violent tur-bulence. The turbulence and gusty winds associated with cumulo-nimbus also extend to the ground for considerable distance around the cloud. The pilot should therefore exercise special caution before attempting a takeoff or landing when a cumulonimbus is near his air-port.

A mammillated* type of cumulus often appears on the lower sur-face of the lateral parts of the anvil or under the low storm collar. This type of "boiling" cloud appears at the beginning of a squall or just before. An example is shown in Figure 7–22.

SPECIAL "CLOUDS"

There are several phenomena related to clouds which are of interest to aviators—tornadoes, waterspouts, contrails, snowstorms, and dust storms.

The Tornado Cloud

The tornado is the briefest of storms, living but an hour or two, but it is also one of the most destructive. It is composed of water droplets which always develop in the lower portion of heavy cumulonimbus clouds. It forms in a funnel-shaped, counterclockwise whirlwind, hang-ing from the thunderstorm base. The whirlwind or cyclone encloses a very sharp barometric depression—as much as 5 in. of mercury. Moist air entrained in this circulation is subject to rapid expansional cooling condensing the vapor. As the tornado reaches the ground, it sucks up dust and debris and causes great damage. Winds in the tornado are estimated to exceed 400 knots with violent shear and turbulence and generally can be expected to destroy any aircraft exposed to it on the ground or in the air.

The diameter of a tornado, or width of the tornado path, is gen-erally less than one-quarter mile but ranges from a few feet to more than a mile. It generally travels in a northeasterly direction at about

---

* From the Latin word *mamma* meaning udder or breast.

35 knots, and its path length on the ground ranges from a few hundred feet to as much as 100 miles.

Tornadoes have occurred in almost every state in the United States, but the Mississippi Valley is the most active region. The atmospheric conditions necessary for the formation of tornadoes are not known precisely. They are usually expected, however, when an active cold front east of the Rocky Mountains separates polar air from maritime tropical air, a moist tongue of warm air is flowing northward from the Gulf of Mexico, and a dry high-level jet stream from the west crosses this moist flow at nearly right angles.

Needless to say, pilots should avoid tornadoes. Even the airlines will usually cancel all night flights through tornado forecast areas.

The Waterspout

The waterspout is similar in nature to the tornado, though usually far less violent and not as frequent because of the ocean's moderating influence on instability, which prohibits large air-sea temperature differences. It usually occurs in tropical and semitropical latitudes, although spouts have been observed at sea as far north as 40° north.

Most waterspouts originate from the 1500-ft to 2000-ft base of a stratocumulus or cumulonimbus cloud. Beneath such clouds, a whirl will start from the cloud base, then extend downward in an irregular funnel shape to about the 1500-ft level. From this point, an unsteady, whirling, tube-shaped cloud 20 to 40 ft in diameter will descend toward the surface of the ocean.

Directly under the end of the tube, the water becomes violently disturbed and begins to whirl in a counterclockwise direction. From the whirling area, which has a diameter of about 75 ft, light clouds of vapor, known as cascade, rise to meet the snout of the descending tube. The cloud tube and the rising vapor rotate around a hollow core.

Some waterspouts never make a complete junction with the surface of the water. In those incomplete spouts, the tubes extend to between 500 and 1000 ft of the water, and then withdraw into the clouds above.

The complete waterspouts remain whole from five to ten minutes, and then break at an altitude of about 250 ft, the upper sections of the tubes being drawn upward, while the lower sections dissipate in the surrounding air.

Contrails

Contrails are man-made clouds which form in the wake of an aircraft when the atmosphere at flying level is sufficiently cold and humid. The main factor in the formation of contrails is the sudden cooling of exhaust gases which have a high water-vapor content as a result of the combustion of fuel. When formed, contrails have the appearance of brilliant white streaks. Usually contrails are short-lived, although they may persist for several hours.

If for some reason a pilot desires not to make contrails, he can usually do so by either climbing or descending. By so doing, he gets out of the band of air in which the conditions of moisture and temperature are suitable for contrails.

Storms of Blowing Snow

Areas of heavy precipitating snow obviously can be serious obstructions to visibility just as heavy rain can be. In the wake of snowy precipitation, however, dry, cold, powdery snow often presents a serious visibility problem to pilots. This type of surface snow cover can be caught up in winds and form a dense layer as much as a few hundred feet thick. Such a condition is really a hazard only for landings and takeoffs, but even a 10-ft to 20-ft layer of thick, blowing snow over the runway can be dangerous.

Dust Storms

Dust from the surface of dry, powdery soil is also caught up in the wind. In the midwestern plains these dust storms create a visibility hazard up to 10,000 ft thick. The stronger the wind and the more unstable the air, the thicker these dust layers become. Stable air tends to keep the dust lower and more concentrated, presenting special problems in approach to landing.

The frequency and extent of the plains-states dust storms vary with the length and intensity of drought. Two or three years of lower-than-average rainfall in this area sometimes gives rise to vast extents of thick layers of dust covering several states. Shorter periods of more localized rainfall deficiency are naturally favorable to smaller, more localized dust storms, but these can be just as serious as the vast dust storms to pilots operating at airports covered by them.

FLIGHT SIGNIFICANCE OF CLOUDS TO THE PILOT

As mentioned earlier, the pilot who knows the clouds, their types and names, can forecast with some accuracy the type of weather he is likely to encounter during flight. Here are how the clouds translate to a pilot in terms of flight phenomena:

Cirriform   These clouds, being composed entirely of ice crystals, present no serious icing hazard, although the crystals, being extremely cold and dry, may produce precipitation static which will impair reception on low-frequency radio equipment.

Stratiform   Produced under stable conditions, these clouds indicate smooth flight conditions. If encountered in icing conditions, ice would be predominantly rime. Precipitation is of a steady, continuous nature.

Cumuliform   A product of instability, they are an indication of turbulence. Clear, or glaze, ice would predominate in icing conditions. Precipitation is showery, frequently heavy in intensity.

Cirrus   The light, wispy or feathery cirrus, when detached from other cloud formations, is a good indicator of fair weather. If they are drawn out in long bands advancing across the sky, cirri are excellent indicators of strong winds aloft blowing in the direction of the advancing cloud, that is, parallel to the bands. Thus, cirrus can indicate the approach of changing weather.

Cirrostratus   If it appears within a few hours after cirrus in midlatitudes, there is an 80% probability of rain within the next 24 hours. (In the tropics, however, you will see cirrostratus quite often and no rain follows.)

Cirrocumulus   Turbulence exists at the cloud level and probable storm activity follows due to an unstable air mass.

Altostratus   If it follows cirrus and cirrostratus or cirrocumulus by a few hours, a 90% probability for rain within 9 to 12 hours exists. Therefore, altostratus usually indicates the proximity of unfavorable flying weather and precipitation. Light rain or heavy snow may often fall from altostratus clouds.

Altocumulus   Extreme turbulence in the intermediate altitudes (6500 to 20,000 ft) is indicated by towering altocumulus. High winds blowing across mountain ranges may be indicated by the lens- or cigar-shaped altocumulus lenticularis on the lee side of the range at the crest of an area of updrafts.

Stratus   The formation may lower to fog after sunset due to the absence of surface heating and possible decrease in winds. Precipitation is normally in the form of light drizzle, small, closely spaced water droplets. Within a thin layer, the upper surface indicates contours of the surface beneath—hills and mountains as elevated areas, rivers and valleys as depressions.

Stratocumulus   Turbulence and wind of moderate to strong intensity may form stratocumulus from stratus. Wind direction is perpendicular to or across the rolls. Updrafts are under the thicker portions. Above the cloud layer, the air is smooth, but it is turbulent below and within the layer. Pilots flying light aircraft should be cautious beneath these clouds.

Cumulus   Cumuli of little vertical development are good indicators of fair weather.

Cumulonimbus   A veritable "weather factory," the cumulonimbus is one of the worst hazards the aviator may ever encounter in flight. Besides the extreme turbulence, particularly in the upper portions, there are icing, heavy rain, snow, hail, lightning and gusty winds and sharply reduced visibilities at the surface. Cumulonimbus may be isolated or in a line sometimes several hundred miles in length. Watch out for hail underneath well-developed cumulonimbus.

A special cloud of interest to pilots is the so-called *roll* or *rotor cloud* (or squall line) which forms on the lee side of mountains and in the roll cloud in the leading edge of thunderstorms. These are *very* turbulent and dangerous to low-flying and light aircraft.

STABILITY AND CLOUDS

The types of clouds we have been describing, and the weather that we experience in flying, depend on whether our atmosphere is "stable" or "unstable." If our atmosphere is *stable,* clouds are likely to be layer-like and the formation of fog is a possibility. On the other hand, if our atmosphere is *unstable,* we are likely to encounter clouds of strong vertical development, such as thunderstorms. Hence, an understanding of the atmosphere's stability is a big help for pilots to understand how weather and clouds are produced.

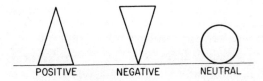

Figure 7–23 Stability.

First of all, let us understand the terms which meteorologists use about stability:

Positive (or absolute) stability

Neutral stability

Negative, or absolute *in*stability

A cone and a sphere will illustrate these terms. (See Figure 7–23.) If the cone is sitting on its base, it is said to have positive stability. If you move the point, it will tend to return to its former position. On the other hand, if the cone is balanced on its point, it is said to have negative stability (or absolute instability), for even the slightest force will make the cone move further and faster away from its original position. The sphere, on the other hand, is said to be neutral. If it is moved, it will assume a new and stable position after the force is removed.

So it is with our atmosphere. As we have learned in Chapter 6, the circulation of the atmosphere tends to be horizontal. If the atmosphere is *stable*, it resists any upward or downward displacement and tends to return quickly to a normal horizontal flow. On the other hand, an *unstable* atmosphere will permit upward and downward air movements to grow, resulting in turbulence, thunderstorm activity, heavy precipitation, and hail. Stability, therefore, is atmospheric resistance to vertical motion.

What determines whether air is "stable" or "unstable"? Stability depends on the vertical distribution of the air's weight and moisture at a particular time in relation to its surroundings. If air is warmer than its surroundings, it will rise, just like a heated balloon rises. If it is colder than its surroundings, it will sink. If it is similar to its surroundings, it will remain in its position.

Vertical currents in the atmosphere can be induced by (1) thermal convection, (2) mechanical convection, either orographic lifting or frontal lifting, and by (3) convergence.

Thermal Convection

Air which is resting on the earth's surface can be set in motion by surface heating which is most noticeable during the afternoon. Vertical currents may also result when cool air moves over a warm surface and is heated by contact with the warm surface.

Mechanical Convection

Air may be lifted by one of two physical barriers:

(1) *Orographic lifting*   As air moves across the earth's surface, it is frequently forced to rise over hill or mountain barriers. (See Figure 7–24.) Even mechanical turbulence caused by the passage of air at high speeds over rough terrain features such as buildings, ridges, gulleys, and similar obstructions may extend to altitudes of 10,000 to 14,000 ft in clear air.

(2) *Frontal lifting*   Where air mass boundaries (fronts) exist, the warmer air is forced to rise over the denser cold air mass.

Convergence

Convergence is the inflow of air into an area. It may result from different wind velocities in relatively straight-line flow or from the flow of wind across the isobars at the surface into regions of low barometric

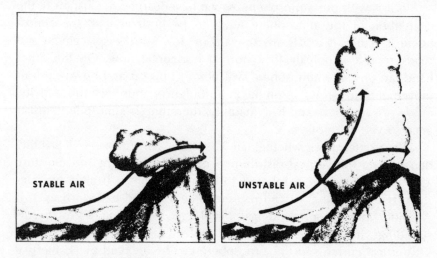

Figure 7–24   The Effect of Stability on Cloud Structure.

pressure. As the air rushes into the cyclonic center from all sides, it rises in great volumes to compensate for the convergent action.

Atmospheric stability is dependent on what the meteorologists call *lapse rates*—the *rate* of temperature decrease. As we learned in Chapter 3, the temperature decreases *on the average* at 2°C. per 1000 ft (called the *standard* lapse rate)—but considerable departures from this average are frequent, depending on whether the air is saturated or unsaturated.

Dry Adiabatic Lapse Rate

When dry or unsaturated air rises, its temperature decreases at the rate of 3°C. per 1000 ft, whether the air is forced upward as a result of being heated from below, or through orographic lifting up a mountain. This is known as the *dry adiabatic lapse rate.*

When dry air rises, it expands. Heat, a form of energy, is consumed as the dry air expands, thus removing heat and causing cooling. This is called an *adiabatic process,* the word "adiabatic" meaning that the temperature change takes place without adding or taking away heat. Warming by contraction is the reverse of cooling by expansion, and is also an adiabatic process.

Moist Adiabatic Lapse Rate

When wet or saturated air rises or is forced upward, condensation occurs and the heat released in the condensation process is absorbed by the air. This causes moist air to cool more slowly than drier air. Thus, we have a *moist adiabatic lapse rate* which can vary from 1.1°C to 2.8°C per 1000 ft.

Absolute Stability

By comparing the actual lapse rates of various layers with the dry and moist adiabatic lapse rates, the degree of stability of the atmosphere can be estimated. When the actual lapse rate is *less* than the moist adiabatic lapse rate, the air is said to be *absolutely* stable. A parcel of absolutely stable air which is lifted becomes cooler than the surrounding air and sinks back to its original position as soon as the lifting force is removed. Similarly, if forced to descend, it becomes warmer than the surrounding air and, like a cork in water, rises to its original position upon removal of the outside force.

Absolute Instability

When the actual lapse rate in a layer of air is *greater* than the dry adiabatic lapse rate, that air is absolutely *unstable*, regardless of the amount of moisture it contains. The plot of the actual lapse rate lies to the left of the dry adiabat. A parcel of air lifted adiabatically even slightly will at once be warmer than its surroundings and, like a hot-air balloon, will be forced to rise rapidly.

Conditional Stability

When the actual lapse rate lies between the dry and moist adiabats, the air is conditionally stable. If the air is saturated, it will be unstable; if unsaturated, it will be stable. In other words, whether the air is stable or unstable depends upon the amount of moisture it contains. The standard lapse rate lies between the dry and moist adiabats, indiacting that, on the average, air is conditionally stable.

Figure 7–25 shows these three conditions.

In conclusion, there are four possibilities in respect to stability:

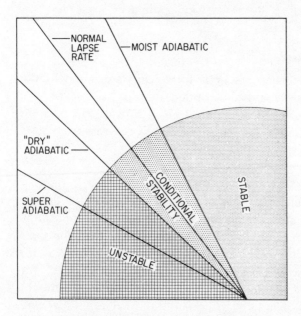

Figure 7–25 The Degree of Air Stability in Relation to the Rate at Which Temperature Changes with Height.

(1) If the lapse rate is less than the moist adiabatic rate, then a parcel of rising air will be *absolutely stable.*
(2) If the lapse rate is greater than the dry adiabatic rate, then a parcel of rising air will be *absolutely unstable.*
(3) If a lapse rate equals the dry adiabatic rate, then a parcel of unsaturated air will be neutral, and if the lapse rate equals the moist rate, then a parcel of saturated air will be *neutral.*
(4) If the actual lapse rate falls between the dry and the moist adiabatic rates, then the parcel will be stable until saturation and then may become unstable after saturation (conditional instability).

After an upper-air sounding is taken, a stability index is computed. There are several types of indices. One is called a Showalter index, which shows the atmospheric potential for the production of thunderstorms. The forecasting rule is that if the difference is +3 or less, showers are probable and thunderstorms are possible; if it is 1 to −2, thunderstorms are probable; if it is −3 or less, severe thunderstorms; −6 or less, tornado conditions.

The weather analysis center in Washington plots the stability indices from all the stations in the United States and makes an analysis of them which is transmitted on the facsimile circuit.

HIGH CLOUDS (Above 20,000 ft)	
1	thin cirrus
2	thick cirrus
3	cirrus or anvil cloud
4	tufted cirrus increasing
5	cirrus or cirrostratus
6	cirrus and cirrostratus
7	cirrostratus covering entire sky
8	cirrostratus partial sky cover
9	cirrocumulus and cirrus

WHAT EVERY AVIATOR SHOULD KNOW ABOUT CLOUDS

The symbols shown below are used by weathermen all over the world.

Ci "FEATHERY CLOUDS"

Often seen during fair weather.

At times serve as first visible indication of approaching storm.

Cirrus clouds are observed at very great altitudes and owe their fibrous and feathery appearance to the fact that they are composed entirely of ice crystals. Although the word "cirrus" derives from the Latin for "curl" or "lock," the clouds are found in varied forms including curved wisps, featherlike plumes, isolated tufts, and thin lines. Because of their height, they color before other clouds at sunrise and remain lighted after sunset.

Cc "MACKERELS' SCALES"

Look for wind and rain if they change to cirrostratus and lower thicker clouds

Cirrocumulus are similar to cirrus clouds but contain globular cotton-like masses arranged in groups .or lines which at times give them the appearance of rippled sand on the seashore. One form of cirrocumulus commonly known as the "mackerel sky" because of the way in which the pattern resembles the scales on the back of a mackerel. The harder and grayer variety often indicate foul weather may follow.

Cs "HALO PRODUCING"

Bad weather approaching if these clouds thicken and change to altostratus.

Cirrostratus covers the sky with a thin whitish veil. The cloud layer is not sufficiently dense to obscure or blur the outlines of the sun or moon. However, the ice crystals of which the cloud is composed, refract the light which passes through them in such a way that a ring known as a "halo" forms around the sun or moon. Cirrostratus clouds which follow after cirrus may be an indication of approach of low-pressure area.

Ci "MARES' TAILS"

This type appearing after cirrus and followed by thickening lower clouds, increases probability of rain within 24 hours.

Cirrus and cirrostratus. "Mare's tails" is the popular name given to well-defined cirrus clouds that thicken into cirrostratus, and then gradually lowering into water droplet altostratus. The clouds may resemble a mare's tail and may often be the forerunner of a storm as indicated in the old rhyme: "Mackerel sky and mare's tails, make tall ships carry low sails." The more brush-like the cirrus, the stronger the wind at that level.

(Adapted from "All Hands" for 1965)

MIDDLE CLOUDS (6500 ft to 20,000 ft)		
1	∠	thin altostratus
2	⫫	thick altostratus or nimbostratus
3	﹈	altocumulus
4	﹖	altocumulus in small patches
5	⅛	altocumulus
6	χ	altocumulus
7	ß	altocumulus with altostratus
8	⋔	altocumulus in tufts
9	﹖	irregular altocumulus

Often short-lived, making only a brief appearance.

Frequently precede thunderstorms.

"CASTLES IN THE AIR"

Altocumulus. These "castles in the air" are visible proof of the great altitude to which rising currents in the atmosphere often extend. Generally arranged in a line and resting on one horizontal base, they give the impression of turrets on a castle. These turreted tops look like miniature cumulus clouds and possess considerable depth as well great length. These clouds usually indicate a change to chaotic, and thundery skies.

Continuous rain or snow may follow thickening altostratus in a few hours.

"THICK GRAY CURTAIN"

Altostratus clouds have the appearance of a gray or bluish, fibrous veil or sheet which is sufficiently dense so that the sun and moon generally appear as they would through ground glass. There is no "halo" as usually seen through cirrostratus but a similar phenomenon called a "corona" may be observed. The low ragged "scud" or NIMBOSTRATUS "rain clouds" that form under altostratus clouds grow denser and lower as rain falls.

If this formation precedes lower cumulus clouds look for thundery weather.

"SHEEP BACKS"

Altocumulus clouds (known as "sheep backs") are a layer of large, ball-like masses often so close together that the edges touch. They are often mistaken for an unbroken layer of stratocumulus. While the balls or patches may vary in thickness and color—from dazzling white to dark gray—they are more or less regularly arranged and distinct. They differ from cirrocumulus cloudlets in that they show distinct shadowed portions.

These rolls stretch to the horizon and move at right angles to their length.

"LONG ROLLS OR BANDS"

Altocumulus are often in "bands" or "long rolls." This is a form of this cloud type having big roll clouds separated by streaks of blue sky. The rolls appear to be joined together near the horizon because of the effect of perspective. These regular parallel bands of altocumulus differ from the "mackerel sky" in that they are found in larger masses with shadows and are not composed if ice crystals like the higher cirrus forms.

LOW CLOUDS
(Surface to 6500 ft)

1	⌒	cumulus or fair weather
2	⌒	swelling cumulus
3	⌒	cumulonimbus ragged tops
4	⌒	cumulus flattened to stratocumulus
5	⌒	stratus or stratocumulus
6	—	stratus or fractostratus
7	- - -	low broken of bad weather
8	⌒	cumulus and stratocumulus
9	⌒	cumulonimbus

This type generally seen in fine weather.

Turbulence increases as thickness increases.

"WOOLPACK"

Cu

Cumulus clouds are the small, fluffy, "fair weather type." The various types of clouds in the cumulus family are defined according to the extent of their vertical development—the height to which warm moist air is being raised by updrafts within them. It is the presence of these updrafts which makes flying near or in cumulus clouds "bumpy" and sometimes dangerous. Note little vertical development.

This is the signpost of turbulent, bumpy air, with thunder, lightning, snow in upper levels, hail and heavy rain.

"THUNDER HEAD"

Cb

Cumulonimbus "thunderheads" or "showerclouds" heavy masses of clouds rising in mountainous towers to great heights. The upper parts consist of ice crystals and often spread out in the shape of an anvil. The base is horizontal, but as showers occur it lowers and becomes ragged. The anvil of this giant cloud is so high that it can be seen many miles away long before the base becomes visible. A regular "cloud factory."

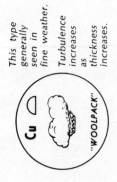

Tail-end of the days cumulus clouds.

Usually a clear night ahead over land.

"FLAT LONG LAYERS"

Sc

Stratocumulus clouds are the final product of daily changes in cumulus clouds. They vary greatly in altitude. At lower levels this type also appears as roll-shaped masses which are soft and gray and can be composed of long parallel rolls. (Such rolls are good indicators of wind direction at their level because they form on crests of atmospheric waves at approximate right angles to the wind producing them.)

Stratus often produce a fine drizzle or mist.

"LAYERS OR SHEETS"

St —

Stratus formations are low horizontal, uniform layers of clouds. Strong winds sometimes break them up into irregular fragments or shreds called FRACTO-STRATUS. A veil of true stratus gives the sky a hazy appearance. Because of their thickness, stratus appear dark to sailors and landmen, but look white to aviators. Clouds of stratus family are called "low stratus" if their base is below 1000 ft and "fog" when on the ground.

8 AIR MASSES

An *air mass* is defined as a widespread body of air, ranging from about 500 to 5000 miles in width and from several thousand feet to several miles deep, with approximately uniform properties of temperature, humidity, and thermal structure (horizontally) which vary only slightly from point to point within the air mass.

GENERATION OF AN AIR MASS

In certain localities of the earth, meteorological conditions are such that the air lying over these areas will have little or no tendency to move toward another area. If such conditions persist for a sufficient period, the overlying air will acquire definite characteristics and properties from the ground up as dictated by the physical and geographical nature of the surface. By the time such a process has been completed, an extensive portion of the atmosphere has become practically homogeneous throughout and its properties have become more or less uniform at each level, fulfilling the air mass definition given above. When this happens, an air mass has been generated.

An ideal air mass is identifiable by two important characteristics: Its humidity and temperature are the same everywhere at each level. Actually, this ideal never occurs, but the variations in these properties are slight and always gradual; there are no abrupt discontinuities within the same air mass.

Conversely, the transition from one air mass to another is generally marked by an abrupt change in various characteristics, and these changes define the boundaries or fronts of the air masses.

CLASSIFICATION OF AIR MASSES BY SOURCE REGIONS

The region where an air mass is generated, and from which it derives its initial properties, is called its *source region*. Various factors affect the nature of air masses, their physical properties, and their internal structure while they lie over the source region. These factors include general latitude and surface nature of their source, their moisture content, stability, their temperature relative to the surface over which they travel, and the length of time the air remains over the source region. The complete classification provides a fairly accurate description of the air mass in its original and its current state. Figure 8–1 shows the source regions and general movement of North American air masses.

Source Regions

The four main source regions of air masses are (1) the equatorial belt of uniform pressure (the doldrums), (2) the subtropical anticyclone belts, (3) anticyclones in high latitudes, and (4) the great arctic and antarctic regions of ice and snow. The middle latitudes are poor source regions. We may thus distinguish the air within such masses as:

Equatorial	(from the doldrums)	warmest
Tropical	(from the subtropical lows [anticyclones])	warm
Arctic	(from the regions of ice and snow)	coldest
Polar	(from the highs [anticyclones] near the poles)	cold

In reality, there is relatively little difference between arctic and polar air and between tropical and equatorial air.

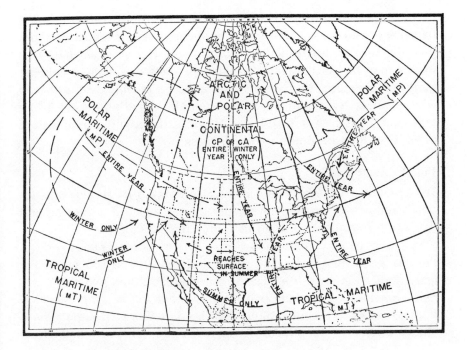

Figure 8–1   General Movement of North American Air Masses.

Surface Nature of Source

The nature of the surface over which air travels will definitely affect the properties of an air mass. The temperature difference between the air and the surface controls the amount of heat absorbed by the air. The air picks up more moisture over the ocean than it does over a land area. Air masses are therefore classified according to the surface nature of their sources, as *maritime* and *continental*.

Location of Source Region

An additional classification, by the geographical name of the source region, further describes the properties of an air mass. Thus, air from the Gulf of Mexico would be warm and moist; "Sahara" air would be hot and dry; polar "Pacific" air would be cold and wet, etc.

Air-Surface Temperature Difference

Finally, air masses are classified according to the relative temperature of their surface layer compared to that of the surface over which they travel. An air mass which is warmer than the surface over which it is traveling is called a *warm air mass;* and, conversely, if the air is colder than the surface, it is called a *cold air mass.* The same air mass as it moves may be, successively, a warm air mass and then a cold air mass, or vice versa. Thus, an air mass at 60°F., moving over an ocean whose surface temperature is 50°F., would be called *warm;* but, when that same mass passes inland over a land surface with a temperature of 70°F., it then becomes a *cold* air mass.

Time En Route

The *amount* of change in the physical properties of an air mass depends upon two things, namely, the rate at which it is moving and the length of time since it left its source region.

So long as an air mass passes over a fairly homogeneous surface, the changes within the mass will be fairly consistent throughout the mass; thus its internal irregularities are gradual. When two air masses meet, however, particularly two from different sources, the transition from one mass to the other along their borders *(fronts)* is often abrupt. These differences in physical properties, especially in temperature, along the border or front between the air masses, produce great contrasts in energy. Along those boundaries, also, traveling lows or depressions develop from the contrasts in energy.

In summary, there are four basic types of air masses: cold dry, cold moist, warm dry, and warm moist. A pilot will hear the weather forecaster refer to them as continental polar, maritime polar, continental tropical, and maritime tropical.

In the Northern Hemisphere, Alaska, Canada, and Siberia are the principal winter source regions for cold, dry air masses; the polar portions of the Atlantic and Pacific Oceans are the main source regions for cold, moist air masses.

The tropical regions of the Atlantic and Pacific Oceans are the main source regions for warm moist air masses, and arid regions of Africa, Asia, and Australia are the principal source regions for warm, dry air masses. On occasion, the southwestern portion of the United States and the northern portion of Mexico become a source region for warm, dry air masses in North America.

Table 8–1, which pilots should know, defines the source regions most used on weather maps in North America and the abbreviations commonly employed to designate them.

Table 8–1

TYPE	SYMBOL[a]	SOURCE REGION
Polar Continental[b]	cPk	Canada, Alaska, Arctic region
	cPw	
Polar Maritime[b]	mPk	Northwestern Atlantic; North Pacific, par-
	mPw	ticularly in vicinity of Aleutian low
Tropical Continental	cTk	Southwestern U. S. and Northern Mexico,
	cTw	in summer only
Tropical Maritime	mTk	Sargasso Sea, Caribbean Sea, Gulf of Mex-
	mTw	ico, subtropical

[a] Explanation of symbols: A is Arctic; T is Tropical; P is Polar; m is maritime; c is continental; k is colder than underlying surface; w is warmer than underlying surface.
[b] The word polar does not mean air around the poles. Air masses in the immediate polar region are designated as *arctic*.

GENERAL CHARACTERISTICS OF AIR MASSES

The following weather conditions are typical of the air masses with which they are identified. Knowledge of these weather characteristics will aid a pilot in predicting the flying conditions likely to be found within any given air mass.

Cold (k type) Maritime Air Mass (mPk)

General characteristics of a cold maritime air mass are:
(1) Cumulus and cumulonimbus-type clouds
(2) Generally good ceilings (except within precipitation areas)
(3) Excellent visibility (except within precipitation areas)
(4) Pronounced air instability (turbulence) in lower levels due to convective currents
(5) Occasional local thunderstorms, heavy showers, hail, or snow flurries

Warm (w type) Maritime Air Mass (mTw)

General characteristics of a warm maritime air mass are:
(1) Stratus and stratocumulus-type clouds and/or fog
(2) Low ceilings (often below 1000 ft)
(3) Poor visibility (since haze, smoke, and dust are held in lower levels)
(4) Smooth, stable air with little or no turbulence
(5) Occasional light continuous drizzle or rain

Continental Air Mass (cP)

Continental air masses are associated with good flying weather; that is, clear skies or scattered high-based cumuliform clouds, unlimited ceilings and visibilities, and little or no precipitation. However, two exceptions are:

(1) Intense surface heating by day may produce strong convection (turbulence) with associated gusts and blowing dust or sand.

(2) Movement of cold, dry air over warm, moist water surfaces may produce dense steam fog and/or low overcast skies with drizzle or snow. If the air continues its movement into mountainous terrain, heavy turbulence, icing, and showers may develop.

These will be discussed more fully below.

COLD MASSES AND THEIR PROPERTIES

Normally, cold masses form in the polar and arctic regions, but in winter they may form over the large land areas in latitudes as low as 25° north. During generation, these masses are cooled from below by contact with the cold surface, and acquire the following properties: low temperature, with consequent low humidity; a gradual lapse rate with marked stability, particularly near the surface.

As the mass moves away from its cold source toward a warmer area, contact with the warmer surface develops instability in its lower layers. This condition spreads upward, destroying any inversions and producing convective currents and (if sufficient moisture is picked up) clouds of the cumulus and cumulonimbus variety.

Cold *continental* and cold *maritime* masses have certain distinct differences by which they may be recognized. All of their typical characteristics as listed in Table 8–2 *may* be found, yet several may be missing, depending upon the degree of modification or "transition" which the mass has experienced in its travel. The following examples illustrate this.

(1) From the broad plains of western Canada in summer, a cold continental air mass moves down over Kansas and Nebraska. As it travels southward, it is heated considerably by the increasingly warmer surface, yet it acquires little moisture.

Table 8-2   Typical Properties of Cold Masses

PROPERTY	CONTINENTAL	MARITIME
Temperature	Increasing	Increasing
Humidity	Fairly constant	Increasing
Lapse rate	Steep, unstable	Steep, unstable
Clouds	Scattered cumulus, occasional cumulonimbus	Plentiful and heavier cumulus and cumulonimbus
Precipitation	Light showers, mostly in the afternoon	Heavy, squally showers, mostly in early morning
Cloudiness	Pronounced daily variation, maximum in the afternoon	Slight daily variation, mostly in early morning
Sky	Considerable bright intervals	Variable between bright and threatening
Cloud base	Considerable; seldom below 2000 ft	Moderate, but seldom below 1000 ft
Visibility	Variable, mostly good, except for dust and smoke	Excellent, between showers

Its humidity and dew point are thus kept low, and its condensation level is kept high. In spite of the instability created by the steep lapse rate in the lower layers, only scattered cumuli will be formed, with occasional cumulonimbus, usually in the afternoon.

(2) Winter air from the cold, dry plains of Texas moves southward out over the warm waters of the Gulf of Mexico. It absorbs a considerable amount of moisture from the ocean's surface, bringing the condensation level lower; the warm contact creates a steep lapse rate in the lower layers of this maritime mass and produces strong convective currents, from which develop cumulus and cumulonimbus clouds; sharp, squally showers follow, with intermittent breaks of clearing weather.

If the mass of (1) above, after being warmed on its journey, had moved out over a cold ocean, a stable state would have been developed in its surface layer, even though its humidity were increased by contact with the water. However, had it moved out over a *warmer* ocean its instability would have been *increased,* with resultant showers of increasing frequency and intensity.

If the mass of (2) above had moved back over a cold surface, whether wet or dry, it would have become stable in the surface layer. Also, at the same time, showers from the cumulonimbus would have reduced the instability by releasing the latent heat of vaporization above the condensation level.

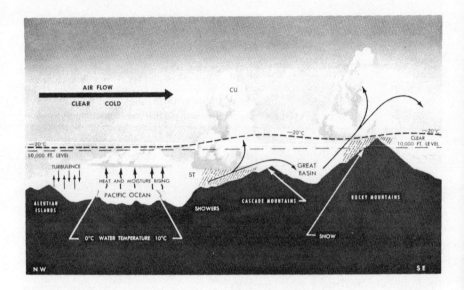

Figure 8–2 Winter Movement of Cold, Moist Air Mass from Gulf of Alaska into Northwest United States.

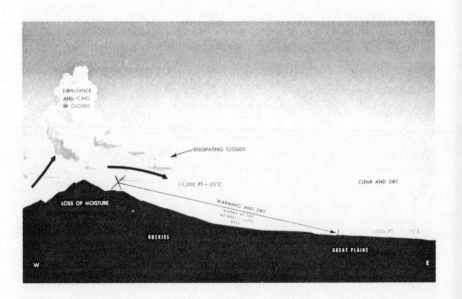

Figure 8–3 Cold, Moist Air Crossing the Rockies.

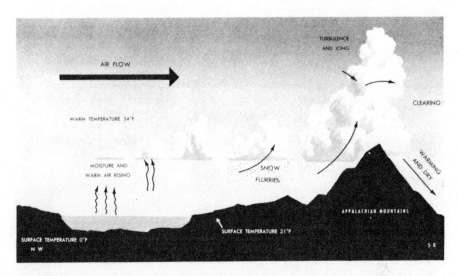

Figure 8–4   Cold, Dry Air Crossing the Great Lakes in Winter.

Figures 8–2, 8–3, and 8–4 show typical movements of cold air masses.

WARM MASSES AND THEIR PROPERTIES

The subtropical anticyclone belts are the most prolific sources of warm air masses, although they may also be formed over the continents in summer under the nearly calm conditions or areas of high pressure or anticyclones.

Warm maritime and warm continental air masses also have distinctive characteristics (see Table 8–3).

A warm maritime mass will be generated with a fairly stable and very humid surface layer. As it travels away from its warm source the increasingly cooler surface serves only to intensify the stability of the lower layer, preventing convectional currents by the strong inversion at the top of the cooled surface layer. Except for the slow cooling incident to outgoing radiation, the air above the inversion will be affected only slightly, although it may become conditionally unstable during the normal daily cycle of variation. Meanwhile, if the surface layer becomes sufficiently cooled so as to drop below its dew point, an advection fog will result. If the wind is sufficiently strong to cause a mixing of the air below the inversion, the fog will spread upward

Table 8-3   Typical Properties of Warm Masses

PROPERTY	CONTINENTAL	MARITIME
Temperature	Decreasing	Decreasing
Humidity	Low; fairly constant	High; increasing
Lapse rate	Conditionally unstable	Conditionally unstable
Clouds	Very few; high	Hazy, stratus-type or fog
Precipitation	Little, if any	Steady, light
Cloudiness	Clear, or a few scattered	Variable to overcast
Visibility	Good	Poor, hazy

to the altitude of the inversion. Mechanical turbulence, however, may heat the air closest to the surface enough to dissipate the lowest part of the fog, resulting in a layer of stratus close to the ground or water. At sea this is almost always the case, so that in a fog, a ship's lookout stationed low in the "eyes" of the ship can often see much farther ahead than a lookout stationed aloft.

Warm continental masses form over dry land surfaces under stagnant or quasi-stagnant conditions during the summer, particularly in the subtropical anticyclone belts. With the exception of those generated over desert regions (such as the Sahara and Arizona deserts) such masses are conditionally unstable. As they travel toward cooler areas, the stability of the lower layer is increased by cooling from below and the formation of an inversion. The air above that, however, changes but little because of the inversion and tends to remain con-

Figure 8-5   Warm, Moist Air Moving over Warmer Surface.

Figure 8–6 Warm, Moist Air Moving over Colder Surface.

ditionally unstable. The mass thus acquires the typical properties of a warm mass below the inversion and those of a cold mass above. Warm air masses generated in the desert regions acquire marked instability by their contact with the hot sands. The convective currents resulting produce "dust devils" or "whirling dervishes," small but rather violent whirling dust clouds that travel along the surface, or, in extreme cases, terrific sandstorms covering extensive areas. As these masses move into contact with cooler surfaces the instability decreases and they gradually become stable in the surface layer. On account of the extremely low humidity of the entire mass even to great altitudes, few if any clouds will form.

Figures 8–5 and 8–6 show typical movements of warm air masses.

SUMMARY

This chapter has described the primary characteristics of the various air masses which influence the weather and flying conditions in the United States. The pilot will encounter many variations of these generalized conditions. To determine the specific conditions for flight within any air mass on a specific day, consult the current weather reports and charts, and discuss the matter with a forecaster at the time of the weather briefing.

Remember these points:

- The advection of cold air over a warmer surface gradually increases the temperature of the air next to the ground and makes it unstable.
- The advection of warm air over a colder surface decreases the temperature of the air next to the ground and makes it stable.
- Ascending air currents (convection) are produced by the heat the air receives while in contact with a hot surface (conduction), by the air being forced over mountains (orographic lifting), or by warm air being forced over cold air (frontal lifting).
- The relative humidity of air increases as the air expands and cools in ascending currents. When the lifting and resulting expansion and cooling are extensive enough to produce saturation, clouds and turbulence are the usual result.
- The relative humidity of air decreases as the air contracts and warms in descending air currents. For this reason, clouds generally dissipate in descending air currents.
- Turbulence, vertical air currents, cumuliform clouds, showery precipitation, and good surface visibility (except in showers or dust) are normally associated with unstable air.
- Smooth flying weather, stratiform clouds, and fair to poor surface visibilities are normally associated with stable air.
- Air masses acquire water vapor by evaporation from underlying water masses or from precipitation falling through the air mass (providing the water temperature exceeds the dew point).
- The water-vapor content of an air mass is reduced by the formation of clouds and precipitation on the windward side of mountains over which air is flowing (orographic lifting). As the air descends the leeward side, it is heated and consequently dried.
- Mountains, valleys, and water masses modify the temperature and humidity of the air mass over a given locality.
- There is a daily cyclical variation of temperature in the surface layers of an air mass. Minimum temperatures normally occur near daybreak. The temperature then steadily rises and reaches a maximum value between 1400 and 1600 local time. A steady decrease in temperature (nocturnal cooling) then takes place during the night and early morning.

9 FRONTS

GENERAL

Since the advent of television, and daily and nightly weather programs, words like "high" and "low" and "front" have become common in our everyday vocabulary.

Actually, the weather word "front" was coined in World War I, taken from the "fronts" of the battlefields of Europe. This usage was initiated by a Norwegian father-son team of meteorologists named Bjerknes who developed the *polar* or *wave theory of cyclone development*. In their study of air masses, the Bjerkneses visualized that a cold mass of air and a warm mass of air, being very different and foreign to one another, would always clash and "fight," and that their line of battle would resemble the battlefields of the Western "front."

Thus, *front* or *frontal zone* is defined as a relatively narrow belt, or, more correctly, an interface or

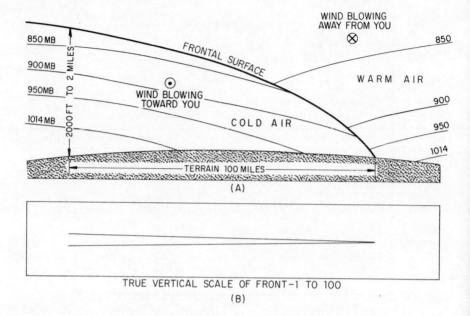

Figure 9–1 (A) Cross Section of a Front. (B) True Vertical Scale of a Front.

transition zone, marking the boundary between two different air masses.

Fronts have sloping boundaries, and, while comparatively narrow, the width across a frontal zone can vary from 3 to 50 miles. In a frontal zone, the air cannot be identified with either air mass since it is a mixture of both. Actual slopes* are much more knife-like than the illustrations in this chapter indicate. (See Figures 9–1A and 9–1B.)

Fronts are of great importance to pilots because of the many hazards to aviation which accompany them. They are usually not recognizable, however, above 20,000 ft. Pilots should remember that fronts are like people—no two are alike.

There are several types of fronts with which the aviator should be familiar, and which will be discussed in this chapter: the polar front, the cold front (both slow-moving and fast-moving), the warm front, the stationary front, and the occluded front (both warm and cold).

---

* An average frontal slope is between 1 in 50 to 1 in 300. A 1 in 100 slope, for example, means that 100 miles inside the front, the front will extend one mile above the ground.

FRONTOGENESIS AND FRONTOLYSIS

How do fronts begin? How do they develop? This question of the formation, location, development, and dissipation of fronts is of the greatest importance to the meteorologist as well as the aviator.

Fronts are spawned by the air masses. Each air mass is a storehouse of tremendous energy; the temperature difference between two air masses is the driving force which releases this potential energy and converts it to kinetic energy. The greater the temperature difference, the more rapidly is the energy released and the more violent is the resulting reaction. This reaction, of course, occurs where two different air masses clash, resulting in moving fronts and storms.

The birth of new fronts, the strengthening or regeneration of old, weak, and decaying fronts is called *frontogenesis,* or frontal development. The opposite process, the decaying and dissipation of existing fronts, is called *frontolysis.*

While much has been learned during the last two decades about how fronts are born and how highs and lows develop, even the most advanced meteorologists are not entirely certain about the processes.

Frontogenesis

Fronts are generated in two ways: (1) by the meeting of air masses of different temperature, and (2) by the creation of a steep temperature gradient along the horizontal, within an air mass, so that a temperature discontinuity occurs. There will also be a density discontinuity.

In the case of two different air masses clashing, the two most active regions for frontogenesis in the Northern Hemisphere in winter are the area between North America and Iceland, and the area between Alaska and Asia. In summertime, the frontogenetic activity is most noticeable between the Bering Sea and central Canada.

Figure 9–2 is a common example of frontogenesis within an air mass. The high-pressure area has been stagnant for several days, and since it covers a large distance from north to south, the air in the north remains colder than the air in the south. Thus, over a period of time, the two "ends" of the high grow to be different. Some of the warmer air in the south is carried north and the middle part of the original high becomes *frontogenetic,* and a temperature difference be-

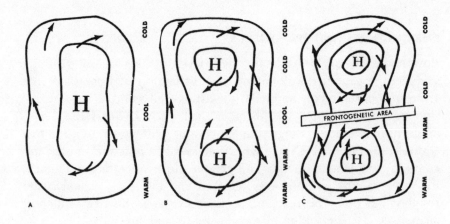

Figure 9-2   Frontogenesis.

comes so marked in this area that two air masses become distinguishable.

Frontolysis

Fronts are destroyed by the release of their contained energy in frontal activity, by the merging and mixing of the air of the two masses along the frontal zone, and by the resultant reduction of the steep temperature gradient until it is no longer a narrow discontinuity but, instead, a broad zone of gradual transition. This process of the weakening or destruction of a front is called *frontolysis*, or front breakdown. Figure 9-3 is an example.

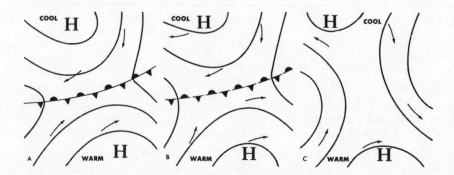

Figure 9-3   Frontolysis.

FRONTAL WAVES

Fronts often have "bends" or "waves" in them. These frontal waves are also the result of the interaction of two air masses. Waves usually form on slow-moving cold fronts or stationary fronts. This process is known as *cyclogenesis,* and while its development is not completely understood, it is generally supposed to be associated with the jet stream.

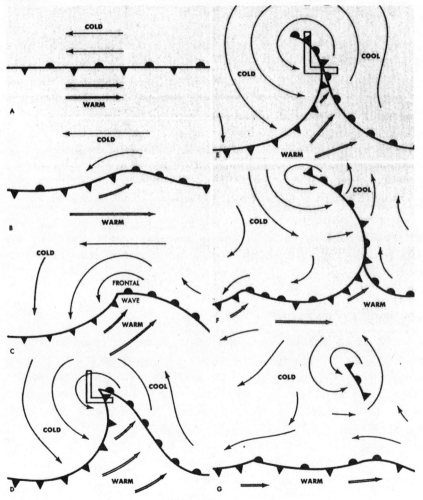

Figure 9–4 The Life Cycle of a Frontal Wave.

In the initial condition in Figure 9–4, the winds on both sides of the front are blowing parallel to it (A). Small disturbances to the steady state of this wind, which are often not obvious on the weather map, as well as perhaps uneven local heating and irregular terrain, may start a wavelike bend in the front (B). If this tendency persists and the wave increases, a counterclockwise circulation is set up. One section of the front begins to move as a warm front, while the adjacent section begins to move as a cold front (C). This produces the frontal wave.

The pressure at the crest of the frontal wave falls and a low-pressure center is formed. The circulation becomes stronger, and the components of the winds perpendicular to the fronts are now strong enough to move the fronts, with the cold front moving faster than the warm front (D). When the cold front catches up with the warm front, the two of them are said to *occlude* (close together), and the process or result is called an *occlusion* (E). This is the time of maximum intensity for the wave cyclone.

As the occlusion continues to extend outward, the cyclonic circulation diminishes in intensity (the low-pressure area weakens) and the frontal movement either slows down (F), or a new frontal wave may form on another portion of the cold front (G).

In other words, the process of cyclogenesis may be strong enough to produce a low cell which will "take off" on its own, or the process may simply develop small weather disturbances along the front.

TERMINOLOGY OF FRONTS

Isobars   On a weather map, lines joining points of equal barometric pressure are called isobars. The system of isobars on a weather map greatly resembles the system of contours on a topographical map. A contour is a line at every point of which the elevation above sea level is the same; at every point on an isobar the barometric pressure is the same. (See Figure 9–5.)

High, or *anticyclone*   A region where the atmospheric pressure is higher than that of other surrounding regions.

Low, depression, or sometimes *cyclone*   A region where the atmospheric pressure is lower than that of other surrounding regions.

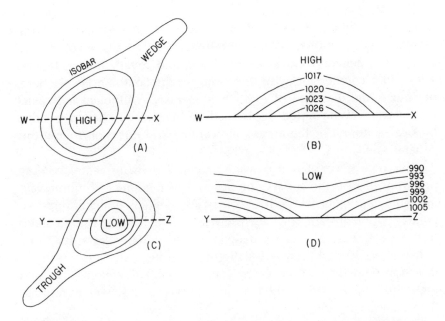

Figure 9–5 A High-pressure Area and a Low-pressure Area, or "Depression," Are Represented by Isobars as a "Hill of Air" and "Valley of Air," Respectively; shown on a horizontal plane at (A) and (C), and in vertical cross section at (B) and (D).

Ridge   An elongated area of relatively high atmospheric pressure.

Trough   An elongated area of relatively low atmospheric pressure.

If we think of isobars as contours depicting the atmospheric topography, the area represented by a *high* represents a hill of air; the area represented by a *low* represents a depression, or valley. This conception is further suggested by the use of the words *high* (for the hill of air) and *low* (for the valley).

On a topographic map, contours close together indicate a steep slope; on a weather map, isobars close together represent a steep hill of air, or steep *gradient*. In contrast with the topographic hill, however, an air hill is *fluid,* and therefore tends to flow down into the valley—the closer the isobars, the faster the rate of flow (wind velocity). As we learned in Chapter 6, surface winds always blow *out* from a *high,* and *in* toward a *low.* If the earth were stationary, this flow might follow the actual direction of the gradient—that is, perpendicular to the isobars—between the two centers. Because of friction, however, the rotation of the earth (the Coriolis force) and related factors, wind in the Northern Hemisphere is always deflected toward the right and approaches the trend of the isobars, blowing more or less along them. Wind, therefore, blows spirally outward from a high in a generally clockwise direction, and spirally inward toward a low, in a counterclockwise direction. (In the Southern Hemisphere, of course, the direction of rotation is reversed.) Although individual exceptions may be noted occasionally on actual weather maps (due to local conditions), the arrows indicating wind direction follow this general rule.

This regular circulation of winds about a pressure center is more pronounced near the surface; on account of the effect of the prevailing winds at higher levels, it becomes less definite as the altitude increases. The highs and lows themselves move across the United States from west to east, at a rate usually in excess of 500 miles a day (from about 20 to 30 mph). Lows usually move in an easterly to northeast direction, and highs in an easterly to southeast direction. Lows and highs in summer are marked by less energy and slower movement than those in winter.

DISTINGUISHING A FRONT

How can a pilot tell when he is crossing a front? There are four indicators:

(1) There will be a temperature change.

(2) There will be a shift in wind direction.

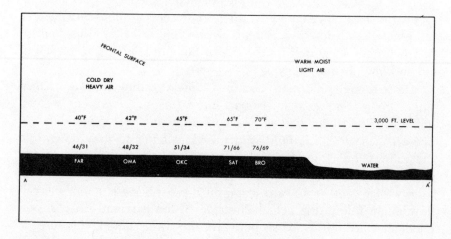

Figure 9-6 Temperature Change at a Front.

(3) There will be a change in the barometer.

(4) The cloud conditions will change.

Take temperature, for example, and look at Figure 9–6. As can be seen, there may be as much as a 20°F. drop in the OAT (outside air temperature) between one side of the front and the other—in just a few miles or minutes. The change in temperature is usually more pronounced at low altitudes than at high altitudes.

While most fronts show several indicators, a few do not.

THE POLAR FRONT

The biggest "battle front" in the atmosphere is the one between the cold air mass of the polar region and the warm air masses of the tropics. These two foes are continually battling each other—cold air pushing south (in the Northern Hemisphere) and warm air moving north, in alternating waves. This zone is called the *polar front*.

The polar front is semipermanent and semicontinuous. Activity along the polar front accounts for a large portion of the weather experienced in the mid-latitudes. Waves form along the polar front (Figure 9–7), and the polar front itself is a dynamic moving boundary. In general, the polar front moves south in the winter and migrates north during the summer. In fact, the polar front may be as far north as 60° north in the summer. Figure 9–8 shows the polar front over the North American continent.

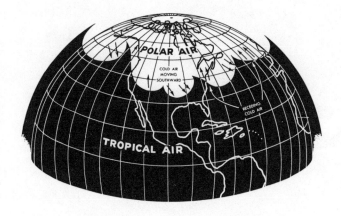

Figure 9–7 The Polar Front.

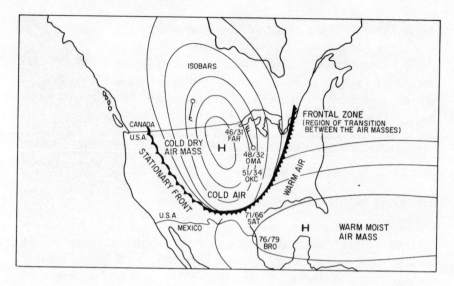

Figure 9–8 The Polar Front in North America.

THE COLD FRONT

Some of the most violent weather a pilot can encounter is associated with the cold front (Figure 9–9A). But, if recognized and proper flight techniques are utilized, cold-front weather will not seriously endanger an aircraft. Figure 9–9B is an unusual picture of a cold front.

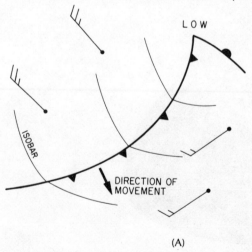

(A)

Figure 9–9 (A) Facsimile Weather Map Presentation of a Cold Front.

(B)

Figure 9–9

(B) An Unusual Picture of a Cold Front (official photograph, Environmental Science Services Administration).

Characteristics of the Cold Front

The cold front is a line of discontinuity where cold air displaces a mass of warm air.

On the surface, a cold front passage is characterized by a temperature decrease, a wind shift, and often gusty winds. Being denser than the warm air, the incoming cold air stays on the surface and, because of surface friction, forms a steep slope. That is, as the cold air moves across the surface of the earth, the air near and on the surface is slowed down. The air aloft, not being subject to the friction, retains its

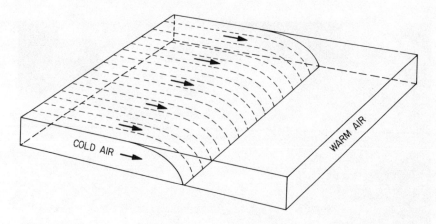

Figure 9–10   The Cold Front.

velocity and tends to catch up to or overrun the air at the surface. (See Figure 9–10.)

Pilots should remember that cold fronts tend to move much faster than warm fronts.

Because of the relatively steep slope of the cold front, the warm air is lifted abruptly. (See Figure 9–11.) This leads to the formation of cumulus-type clouds. This is one of the four identifying features of any front mentioned earlier.

The width of cloud cover associated with a cold front is normally about 50 miles, although it may extend 100 miles or more in some cases (Figure 9–12).

The precipitation associated with a cold front will be of the showery type and can take the form of rain, snow, or hail. With the

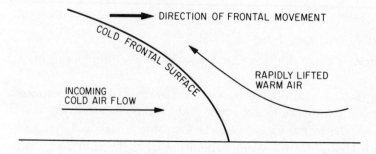

Figure 9–11   Simplified Cold-front Profile.

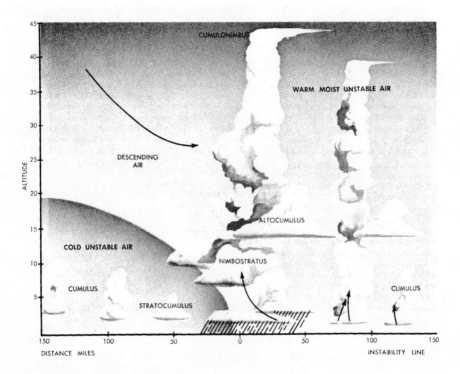

Figure 9–12   A Fast-moving Cold Front.

intense activity usually associated with a cold front, precipitation can be expected throughout the full width of the weather band.

The first sign of an advancing cold front to a person on the ground is a solid line of cumulus clouds approaching from the west or north-west and a drop in barometric pressure. Precipitation will reach the ground very shortly after the first clouds have passed overhead and an abrupt wind shift will occur.

After the cold front passes, the barometric pressure will rise and skies will rapidly clear. The weather behind the cold front will usually be clear skies, or some widely scattered cumulus activity, good visibility, and some turbulence.

There are two main types of cold fronts: slow-moving and fast-moving. A slow-moving cold front, for example, if associated with stable air in both the warm and cold sector, can form nimbostratus clouds over a band of 150 miles or more (see Figure 9–13). (Figure 9–12

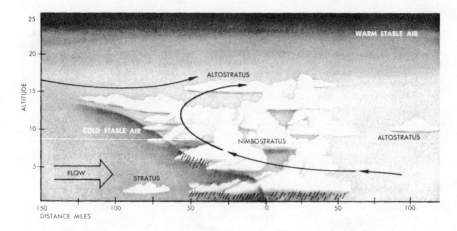

Figure 9–13   A Slow-moving Cold Front.

shows a fast-moving cold front, while Figure 9–13 shows a slow-moving one.)

The greatest hazard in flying through a cold front is the associated turbulence. Icing due to the showery precipitation and large raindrops is usually of the clear or glaze type. Hailstones can develop in the frontal thunderstorms of a cold front.

The most important thing to be remembered about flying through a cold front is to penetrate the front at right angles wherever possible, and to minimize the length of time in it. Flying through frontal weather will be discussed in more detail in Part IV.

In summary, the following characteristics are associated with the approach and passage of the average cold front:

(1) Clouds—generally cumuliform along a narrow line. Specific type, Cb, in a band 50 to 100 miles wide.

(2) Precipitation—showery; may include hail. Icing unlikely, since penetration techniques are based upon avoiding the freezing level.

(3) Flight conditions—generally rough due to turbulence from ground level up to tops of thunderstorms (some recorded in excess of 70,000 ft).

(4) Ceilings and visibilities—generally good except in areas of intense shower activity. Fog is unlikely because of gustiness and high velocity of winds associated with Cb.

(5) Speed and direction of advance—southeasterly at approximately 20 knots.

(6) Slope—100.1.

(7) Wind—shifts from southwesterly to northwesterly with frontal passage.

(8) Temperature—decreases with passage.

(9) Pressure tendency—falls with front's approach, rises after passage.

Squall Lines

One of the weather phenomena most respected by experienced pilots is the squall line of heavy showers and thunderstorms which sometimes precedes a cold front. The air in a rapidly moving cold front sometimes overtakes and passes the surface front. When this happens, a squall line is formed as much as 50 to 100 miles out ahead. The faster the cold front travels, the steeper the slope becomes, and the more violent the thunderstorms produced along the front. The exact cause of these squall lines is not known. One theory is that the strong downdrafts within the thunderstorms carry cold air from aloft to the surface and diverge, forming a wedge of cold air ahead of the front. If the temperature contrast is great enough, the warm air ahead may be forced aloft, creating new thunderstorm activity some distance ahead of the front. This newly generated cloud may project another wedge of cold air ahead, which gives birth to still another new cloud formation. In those cases where this activity takes place all along the front, successive lines of storms may develop and dissipate, progressing ahead of the front itself. The line along which this activity is occurring is termed the instability line, or squall line. The sequence in Figure 9–14A shows this process in action.

The squall line is more violent than the associated cold front, since the moisture and energy in the warm air are released at the squall line. Figure 9–14B is a photograph of a squall line at sea, a rare picture indeed.

The squall line is commonest in spring and early summer in the south central United States. It can bring the most violent weather associated with cold fronts and is one of the conditions usually associated with tornado activity.

When the forecasters are predicting squall lines, therefore, pilots

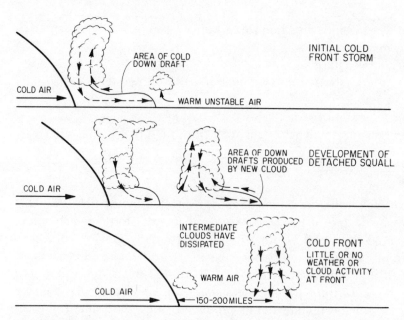

(A) Formation of a Squall Line.

(B) A Squall Line at Sea (official photograph, Environmental Science Services Administration).

Figure 9–14

should be particularly cautious in planning their flights into and around such areas.

THE WARM FRONT

Characteristics of the Warm Front

The warm front is a line of discontinuity where warm air overtakes a mass of retreating cold air. (See Figure 9–15.) Warm fronts move at relatively slow speeds and have gentle slopes. The retreating cold air, being denser, stays on the surface and, because of surface friction, tends to be stretched out ahead of the surface front, forming a rather shallow wedge that lifts the advancing warm air aloft (Figure 9–16).

Because of the gentle lifting action of the warm air, the clouds formed as the warm air rises are of the stratiform type and appear in the following order with the approach of the front: cirrus, cirrostratus, altostratus, and nimbostratus. These clouds can extend as far as 1000 miles ahead of the surface front but normally extend about 500 miles.

At a surface station, the first sign of an approaching warm front (Figure 9–17) would be cirrus clouds thickening to cirrostratus.

As the surface of the upper inversion comes closer to the ground, altostratus would appear, thickening to stratus or nimbostratus. Precipitation usually leads the surface front by 100 to 500 miles, averaging about 300 miles.

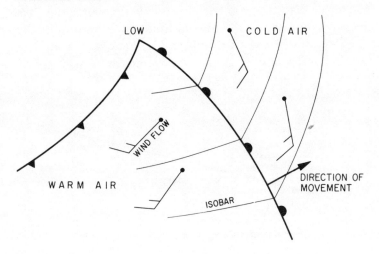

Figure 9–15 Facsimile Weather Map Presentation of a Warm Front.

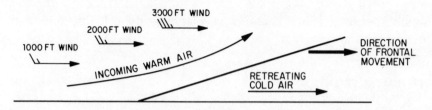

Figure 9–16 Simplified Warm-front Profile.

Most precipitation with a warm front will be of the steady type and can take the form of rain, snow, sleet, or freezing rain, depending on the temperatures associated with the front. Figure 9–18 shows the probable icing in a warm front. When the precipitation is rain, there is usually some evaporation as the warm water falls through the cooler air, saturating the air in the cold sector and forming stratus clouds and fog.

The major hazard which flight through a warm front presents is the poor visibility and fog associated with the stable conditions, making operations around terminals difficult. Also, flight through or along a warm front may require a considerable amount of instrument flying.

Icing which may be encountered in a warm front is due to small raindrops and is usually of the rime type.

If the air in the warm sector is unstable, thunderstorms can form. These have higher bases than most thunderstorms and are not usually quite as violent as the cold-front or squall-line thunderstorms. The foremost danger associated with a warm-front thunderstorm is that the storm is hidden in the stratus clouds and without radar is difficult to avoid. It can give an unwary pilot some trying moments.

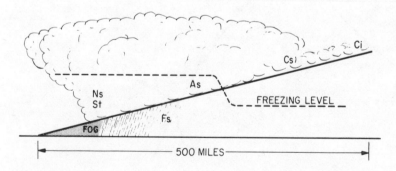

Figure 9–17 Warm-front Weather.

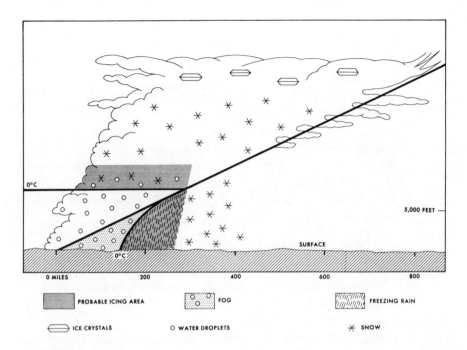

Figure 9–18 Probable Icing Zones in a Warm Front.

The passage of a warm front is marked by an abrupt shift in the wind from southeast to southwest; the barometer, which has been steadily falling, begins to rise and the weather associated with the front dissipates. This normally means clearing, but sometimes if the warm air has a high moisture content, stratocumulus clouds or fog can be formed by the cooling action of the surface that has been cooled by the retreating cold air.

Here are some general rules for a pilot to follow in flying through warm fronts:

(1) Cross the warm frontal areas either above the cloud tops or below the bases to avoid inadvertent entry into the turbulence of hidden thunderstorms which may exist at intermediate flight levels.

(2) If the flight destination is in an area dominated by warm-front weather, an alternate should be selected, either behind the front or as far as practicable ahead of the surface front, to avoid the area of low ceilings and visibility.

(3) Obtain a thorough weather briefing before the flight to estimate the locations of turbulence, thunderstorms, and icing conditions. Keep in touch with the weather ahead by radio while you are en route.

(4) If you encounter an area of freezing rain, immediately climb above the frontal inversion.

(5) Avoid areas of icing in clouds, either by climbing to a higher altitude containing ice crystals and snow, or by descending into warmer air nearer the surface. Since, however, it is possible for two freezing levels to occur with the winter warm front, freezing-level altitudes and flight altitude temperatures should be obtained during the weather briefing.

In conclusion, the following characteristics are associated with the approach and passage of the average warm front:

(1) Clouds—generally stratiform, but some cumuliform may be present. Cloud band—700 miles or more ahead of the surface front.

(2) Precipitation—generally continuous, but may be showery if cumuliform clouds present. Freezing rain probable in winter warm fronts. Precipitation band—100 to 500 miles ahead of surface front.

(3) Flight conditions—generally smooth (except when cumuliform clouds are present). Between layers or in clouds for most of flight.

(4) Ceilings and visibilities—generally poor due to heavy rain, low clouds, and fog (50 to 100 miles ahead of surface front).

(5) Speed and direction of advance—moving northeasterly about 15 knots.

(6) Slope—1:200.

(7) Wind—shifts from southeasterly to southwesterly on the surface (from southerly to westerly above the friction layer).

(8) Temperature change—increase with passage.

(9) Pressure tendency—falling with approach, rising after passage.

OCCLUDED FRONTS

Earlier, the formation of waves or bends along a front was mentioned —bends which lead to the development of an occlusion. If this process

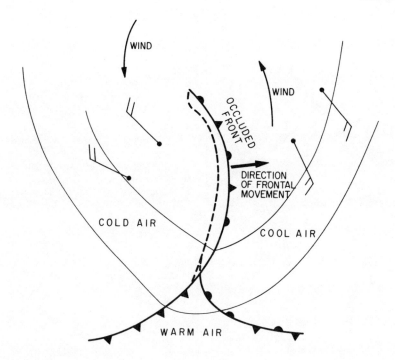

Figure 9–19   Occluded Front.

continues, it becomes an occluded front (Figure 9–19), and the faster-moving cold front overtakes the warm front.

There are two types of occluded fronts, the warm-front occlusion and the cold-front occlusion. The relative temperatures of the cold air ahead of the warm front and the cold air behind the cold front determine what type of occlusion will form.

Cold-Front Occlusion

A cold-front occlusion is formed when the cold air behind the cold front is colder than the cool air ahead of the warm front. Because of its greater density, the air behind the cold front pushes in under the less dense air ahead of the warm front. Because of the steep slope of the cold front, the upper front, shown by a dashed line in Figure 9–20, is only a short distance behind the surface front. For this reason, the upper front in a cold-front-type occlusion is not usually represented on weather maps and is assumed to be in very close proximity to the occluded or surface front.

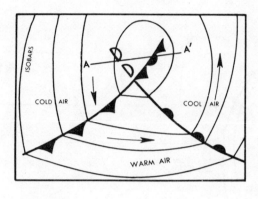

Figure 9–20 A Cold-front Occlusion. The cross section of the cold-front occlusion, (B), occurs at line AA' on the weather map, (A).

(A)

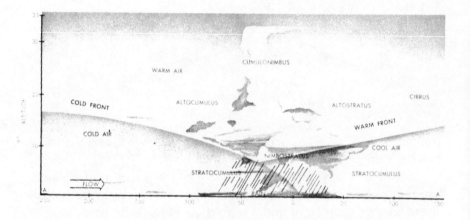

(B)

Cold-front-type occlusions are commonest on the eastern side of continental areas. In the eastern United States, for example, the cold air on both sides comes from the Canadian High. The cold air behind the cold front is moving south from its source region and is relatively cold. The cool air ahead of the warm front has been over warmer terrain for a longer period of time, has modified, and is warmer and thus less dense than the cold air behind the cold front.

Warm-Front Occlusions

Warm-front occlusions normally form in the western part of continents. In the United States, a typical warm-front occlusion would form when a cold front, backed up by cool air from the North Pacific Ocean, moves in over the continent and catches up with a warm front that is preceded by colder continental air.

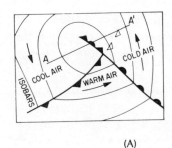

(A)

Figure 9–21 A Warm-front Occlusion. The cross section of the warm-front occlusion, (B), occurs at line *AA'* on the weather map, (A).

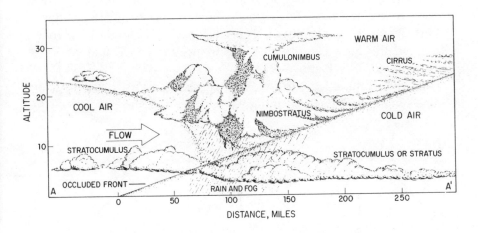

(B)

In this case, the cold air ahead of the warm front is colder than the cool air behind the cold front, and the cold front rides up over the warm front.

The weather associated with an occluded front is a combination of that which occurs when the cold front and warm front exist individually. In the case of the warm- and cold-front occlusions, both will consist of a wide band of clouds and precipitation with a narrow band of turbulence and showers. The warm-front-type occlusion will usually have a broader weather band than the cold-front-type and will form additional clouds in the cold air following the upper front. (See Figure 9–21.)

The greatest hazard encountered while flying through an occluded front is the turbulence within the thunderstorms concealed within the relatively stable layers of the warm front. Its appearance from an aircraft approaching from the east, for example, would be similar to that of a warm front. From the west, the occlusion would resemble a cold front. It is easy to see that the pilot could be misled by the appearance of the front and, unless proper precautions were taken regarding the occlusion, the resulting additional weather caused by the occlusion could result in an aircraft accident report.

Avoiding the Crest

As the occlusion develops from its initial stage, the crest of the surface wave moves to the south. It is in the vicinity of the crest of this surface wave and approximately 50 to 100 miles to the north that the greatest lifting of the warm air occurs and also the area of greatest activity. When flying in the vicinity of an occlusion, flights should be planned to avoid this area because of the extreme turbulence that could be encountered. (See Figure 9–22.)

As the occlusion progresses, the two cold sectors, being over the same surface, will gradually modify until both have the same character-

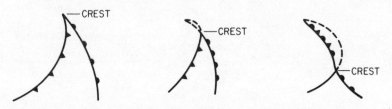

Figure 9–22   Crests of Occluded Fronts.

Figure 9–23   Frontal Dissipation.

istics. When this happens, the surface front loses its identity and the occlusion becomes a temperature inversion with a moderate flow of winds around the low that was previously the crest of the cyclonic wave. (See Figure 9–23.)

Flying the Weather in Occluded Fronts

Here are some tips on flight through occluded fronts:

(1) Plan your flight to avoid the area of severe weather extending 50 to 100 miles along the upper front north of the peak of the warm sector.

(2) Avoid intermediate flight levels where hidden thunderstorms generally occur. In light, private aircraft, a low-level flight below 6000 ft is generally recommended.

(3) While flying at low levels, the occurrence of heavy showers will indicate that stronger turbulence is present in the clouds above.

(4) Low-level flight under the clouds should be avoided where mountainous terrain is obscured by clouds, fog, or precipitation.

QUASI-STATIONARY FRONT

The quasi-stationary front is one which is moving at a speed of five knots or less. For the meteorologist who analyzes the surface map, the quasi-stationary front is one that has not moved appreciably from its position on the last synoptic chart. On occasion, both cold and warm fronts gradually lose speed and show little movement and develop into quasi-stationary fronts.

The weather associated with the quasi-stationary front is dependent upon the moisture, stability, and circulation of the air masses on either side of it. For instance, if warm, stable air is forced by its own circulation over a shallow cold air mass, the weather will be similar to that of a warm front. If the air is unstable, chances are that the more violent type of weather of the cold front will result. Even though cumulus clouds and thunderstorms may occur, they will usually be scattered and not so violent as those associated with an active front.

In summary, the characteristic features of the quasi-stationary front are:

(1) Clouds—similar to the warm front, primarily stratiform but some cumuliform developments with unstable warm air.
(2) Precipitation—generally light and steady except where cumuliform developments appear, then showers as well. Freezing rain and/or freezing drizzle probable during the winter season. Precipitation band: as extensive as low and middle cloud development.
(3) Flight conditions—generally smooth except when cumuliform clouds are present. Low clouds and ceilings in cold-air sector probable.
(4) Speed and direction of advance—less than five knots, described as "drifting."
(5) Winds—gradient winds, as well as isobars, generally parallel to the front. Surface winds within the warm-air sector toward and up the slope at a slight angle.

WEATHER SERVICES FOR PILOTS

10 THE AVIATION WEATHER SYSTEM

In the past quarter-century, a tremendous expansion of our national and international weather service and collection organizations has occurred, which is a measure of the importance that weather plays in aircraft operations, whether civil, commercial, private, or military. Undoubtedly, the growth of long-range, high-altitude, transoceanic, commercial jet aviation has done much to accelerate this expansion, rapid growth, and international cooperation.

The weather knowledge which is available to pilots is derived from a worldwide organization which observes, collects, analyzes, forecasts, and distributes weather information for pilots. This chapter will attempt to explain how this system works, and what part you, the pilot, contribute to it.

In 1870, the U.S. Congress passed a joint resolution establishing a national weather service, and made it a part of the U.S. Army Signal Corps. In 1890, this weather service became a part of the Department of Agriculture and received a new name—the United States Weather Bureau.

In 1938, aviation weather service was made a primary responsibility of the Weather Bureau and in June 1940, the Weather Bureau was transferred to the Department of Commerce. Later, the Civil Aeronautics Authority became the Federal Aviation Agency. In May of 1965, the Weather Bureau, Coast and Geodetic Survey, and the Central Radio Propagation Laboratory of the Bureau of Standards were brought together under an Environmental Science Services Administration (ESSA) which reports to the Secretary of Commerce. Since that time, the two agencies, the Weather Bureau and the Federal Aviation Agency, have continued to expand and cooperate together to service the growing needs of flying.

In 1951, the Flight Advisory Weather Service (FAWS) was organized in the Weather Bureau and 26 forecast centers, later reduced to 20, were established around the United States.

There are several other United States government agencies which cooperate with the Weather Bureau and the FAA in carrying out responsibilities to aviation: Air Force, Navy, National Aeronautics and Space Administration, and Coast Guard.

For many years, pilots obtained all their weather information *before* takeoff. Then came the pilot weather-briefing service where weather observations were made available to pilots over both scheduled radio broadcasts and through ground facilities.

At present, with the spectacular growth of the number of pilots, it has become increasingly difficult for every pilot to get individual weather-briefing service. Under a cooperative arrangement between the Weather Bureau and FAA, the Weather Bureau now trains and certifies FAA Flight Service Specialists as briefers of both current and forecast weather information.

The military weather services of the Air Force, Navy, Coast Guard, and NASA cooperate with the Weather Bureau at all levels. However, the private pilot should understand that the military services have no responsibility for providing aviation weather service, except in emergency situations, since this is the job of the Weather Bureau by law. However, the Weather Bureau and the military weather services exchange information extensively. Weather Bureau processing centers, such as the National Meteorological Center, the Severe Local Storms

Forecast Center, and the National Hurricane Center serve both military and civilian requirements. The services which these centers perform for pilots will be explained below.

While the U.S. aviation weather system was expanding, the worldwide weather organization was also growing. The weather services of Canada and Mexico had always been very cooperative with the U.S. Weather Bureau. But the first worldwide weather organization was the International Meteorological Organization (IMO), founded in 1878. This became the World Meteorological Organization (WMO) in March 1951, and in December of the same year, the WMO became a Specialized Agency of the United Nations.

The WMO is subdivided into six regions—in effect, one for each continent and contiguous waters:

Region I —Africa and contiguous waters

Region II —Asia and contiguous waters

Region III—South America and contiguous waters

Region IV—North America and contiguous waters

Region V —Australia and contiguous waters

Region VI—Europe and contiguous waters

The WMO makes use of more than 8500 weather collection stations all over the world. One of the WMO's most important accomplishments has been to standardize weather instruments and methods of observation, stations, codes, weather reports and to prepare and publish the *International Cloud Atlas*.

THE NATIONAL METEOROLOGICAL CENTER

The main center for the United States weather data is the National Meteorological Center (NMC) of the Weather Bureau, located at Suitland, Maryland, near Washington, D.C.

Throughout its system, NMC receives about 350,000 weather observations each day and prepares a total of 250 analyses and prog charts. Many of these of concern to aviation will be discussed in the next two chapters. Analyses and prog charts for surface and upper levels are transmitted to all U.S. stations, both civil and military, via facsimile network and teletype bulletins. Many of the weather charts distributed by the NMC are prepared automaticaliy. Electronic data processing equipments are capable of plotting limited weather data, analyzing the wind flow pattern, predicting the future pattern, and drawing the analyzed chart and the prognostic chart and automatically transmitting them.

The prognostic charts distributed by NMC are normally good from 18 hours after the observation, but NMC also prepares and distributes 5- to 30-day forecasts.

THE SEVERE LOCAL STORM (SELS) FORECAST CENTER

This center is located in Kansas City, Missouri. It issues warnings of severe thunderstorms and their accompanying hazards including tornadoes, funnel clouds aloft, ¾-in. hail at the surface and aloft, areas of severe or extreme turbulence, and surface wind gusts of 50 knots or more. Such warnings are issued as necessary. In addition, a forecast is issued each morning showing possible locations of severe local storms during the next 24 hours anywhere in the contiguous United States.

Products of the Weather Bureau's severe-storm warning facility are furnished to the military weather services. However, the U.S. Air Force's Air Weather Service also provides hazardous weather warnings which are needed for military operations. The two forecast operations are located in the same building, and the Weather Bureau and the Air Force closely coordinate their efforts.

THE NATIONAL HURRICANE CENTER

This is located at Miami, Florida. All efforts concerning tropical storms or hurricanes which may affect the southern and eastern United States are supervised and coordinated here.

Reconnaissance aircraft of the Weather Bureau, and Air Force and Navy weather reconnaissance squadrons, locate, track, and penetrate hurricanes and other tropical cyclones. These "hurricane hunters" relay information to the National Hurricane Center where it is analyzed and, in conjunction with other information, serves as a basis for forecasts and bulletins. Bulletins and advisories are distributed to Weather Bureau units by teletypewriter and telephone, and to the public and civilian agencies through radio, television, and the press. Issuance of hurricane information to military units is handled through military communications channels by liaison personnel of the Air Force and Navy working with the National Hurricane Center.

The Weather Bureau office in San Francisco, California, serves as the hurricane center for tropical cyclones of all intensities in the eastern North Pacific Ocean and along the west coast of the United States. A service for tropical storms and typhoons in the central and western Pacific is provided through the joint efforts of the Weather Bureau and the military weather services at Hawaii, Guam, and Tokyo. The Navy and Air Force also have forecast centers in Spain and London which provide such services for military installations in the Eastern Atlantic-European area.

THE NATIONAL WEATHER RECORDS CENTER

In order to have the necessary data available from which to prepare climatic studies when they are needed, the Weather Bureau, Air Force, and Navy operate a National Weather Records Center at Asheville, North Carolina. Here weather data are collected, tabulated, and stored from all parts of the world. These data are key-punched and stored. The punch-card system enables the unit to use the IBM (International Business Machine) type of machines for high-speed calculations and tabulation.

HIGH-ALTITUDE FORECAST CENTERS

With the advent of high altitude jet aircraft, the need for high-altitude meteorological information became vital and such a service was begun in 1960. There are seven High-Altitude Forecast Centers in the United States. The main center is co-located with the NMC near Washington, D.C. Its charts are used mainly by the domestic and international commercial airlines. The other six High-Altitude Forecast Centers provide highly specialized forecasts of wind, temperature, significant en route weather, and terminal weather for international flights. These centers are located at Anchorage, Alaska; Honolulu; Miami; New York; and San Francisco, while the sixth is operated in Montreal by the National Meteorological Service of Canada providing information for the North American area. (See Figure 10–1.)

All seven of the High-Altitude Forecast Centers, including the Canadian one, freely exchange forecasts to assure compatibility. The area served by the seven centers extends from Japan and the Philippines eastward into Europe, and from the North Pole to the tropics.

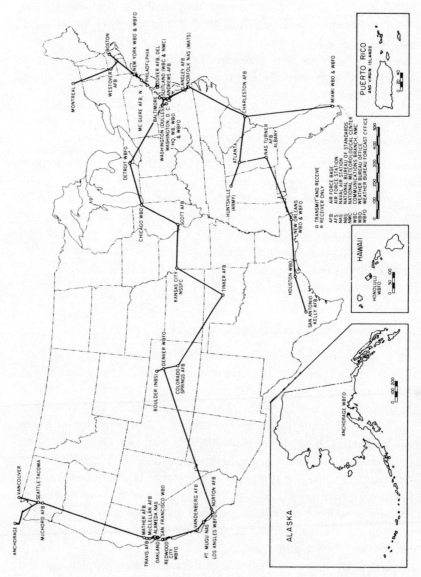

Figure 10-1 High-Altitude Facsimile Network (GF-10200). (Courtesy U.S. Weather Bureau.)

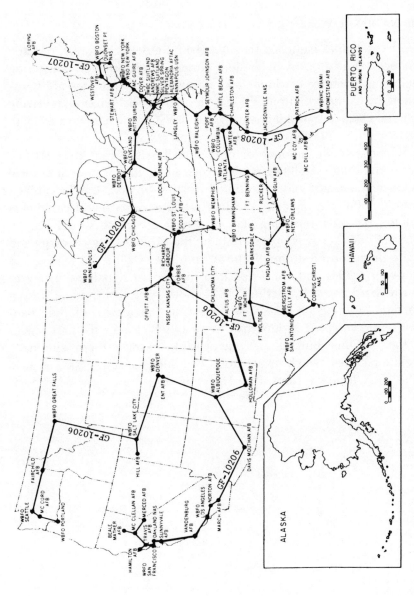

Figure 10-2 Weather Bureau–National Environmental Satellite Center Facsimile Network. (Courtesy U.S. Weather Bureau.)

The High-Altitude Facsimile Network will be described in Chapter 11.

THE NATIONAL WEATHER SATELLITE CENTER

Weather satellites launched from Vandenburg AFB, California, or Cape Kennedy, Florida, carry television cameras and radiation measuring equipment. Pictures of clouds and storms over a large area are transmitted by command from the satellite to special ground receiving stations which retransmit them to the NWSC, which is located with the NMC near Washington, D.C., for processing.

A considerable amount of operationally significant data is now obtained through satellite pictures, especially in the location of tropical cyclones in areas of widely scattered weather observations (ocean and sparsely populated regions). Weather satellites often reveal cloudiness and fog between reporting stations which otherwise would not be known. They also assist in determining the extent of fog coverage and in the detection of storms. Pictures from the satellite are received as television signals and converted into the proper signals necessary to obtain "hard copy" by facsimile (wirephoto). The entire process, from the time the satellite takes the picture to the receipt of the "hard copy," is completed in less than five minutes. Larger forecast offices of the Weather Bureau are equipped to receive these automatic satellite pictures, which reveal cloud conditions as far as 2000 to 3000 miles away in many cases. (See Figure 10–2.)

11 AVIATION WEATHER OBSERVATIONS AND REPORTING

SOURCES OF AVIATION WEATHER INFORMATION

There is an abundance of aviation weather data now available to pilots. As a matter of fact, for the private pilot flying for pleasure from a well-equipped weather airfield, there is far more weather information than is needed for a local or short-range navigational flight.

Where does all this information come from, and how does it reach pilots? There are eleven principal sources: (1) surface observations taken at hundreds of airports, ground stations, and military airfields, (2) pilot balloon observations, (3) radar observations, (4) radiosonde observations, (5) rawinsonde observations, (6) weather reconnaissance flights, (7) meteorological rocket soundings, (8) satellite observations, (9) ships-at-sea reports, (10) automatic meteorological observation stations called *AMOS* (at sea and in isolated land stations), and (11) last, but not least, pilot reports.

Some of these sources are self-explanatory, but others need explanation.

Radar Observations

At many Weather Bureau stations and most military airfields, observations are taken by special weather radar. These are called *RAREPS* and afford a continuous presentation of significant cloud and precipitation patterns in three dimensions. This special weather radar is adjusted to a wavelength that gives the best signal return, considering attenuation from water droplets and other precipitation particles. (This is in contrast to the long-range radar used by air traffic control facilities which employs a wavelength and other devices to *minimize* the signal return from water droplets and other precipitation particles.) Water droplets reflect transmitted radio waves and produce "echoes" which can be seen on the radar scope. The strength of these echoes is evaluated to determine intensity. The size and number of water droplets, as well as the distance to the storm, determine the strength of the return echo.

Weather radar can detect and track severe storms such as thunderstorms, squall lines, tornadoes, and hurricanes.

Radar echoes that are classified as "heavy" or "very heavy" in intensity usually indicate a storm with severe or extreme turbulence, hail, and heavy icing conditions. Echoes of a "light" or "very light" intensity are indicative of snow, light rain, or possibly very heavy drizzle. The strength of the echoes is evaluated by the specialist who correlates the radar presentation with surface reports, pilot reports, and the other sources described below.

There are more than 100 radar-equipped stations in the United States, most of them located east of the Rockies since this is the area where severe storms are most frequent. These stations are grouped into three Radar Warning Circuits. The radars have a range of 250 nautical miles. (See Figures 11–1A and 11–1B.) RAREPS are made every hour by the stations east of the Rockies. A summary of these reports is then made by the Radar Analysis and Development Unit at Kansas City, Missouri. Facsimile charts give the storm's intensity, the type of precipitation, the distance of the storm from known locations, the maximum height, and its expected movement and changes the next few hours. These charts are transmitted eight times daily. Radar information is also promulgated on the teletype circuits described later

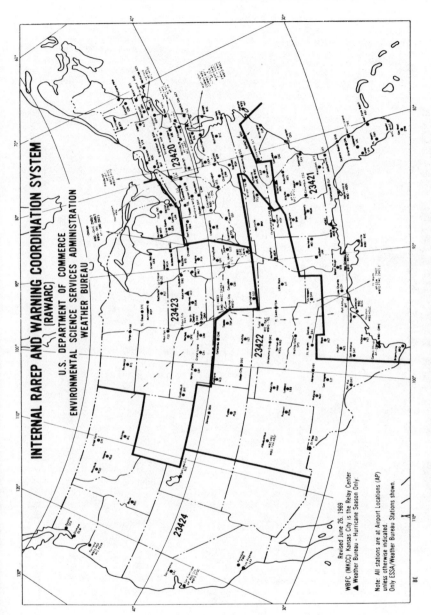

Figure 11-1 (A) Radar Warning Teletypewriter System.

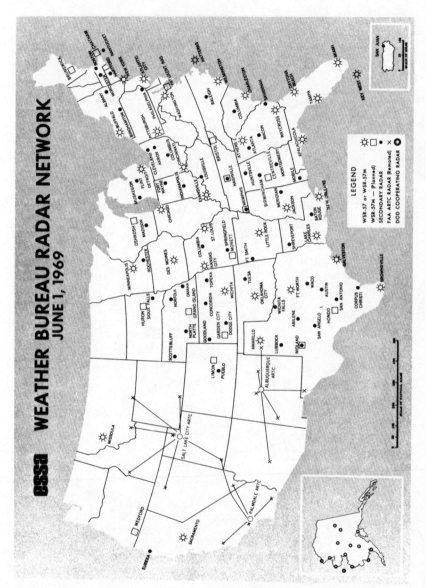

Figure 11-1 (B) National Weather Radar Reporting Network. (Courtesy U.S. Weather Bureau.)

in this chapter. (See Chapter 12 for an explanation of the radar summary charts and the symbols used on them. For the use of airborne radar, see Chapter 18.)

Pilot Balloons

A pilot balloon is a freely floating, 30-in.-diameter balloon filled with either helium or hydrogen which is allowed to escape and ascend. It carries no weather instruments. Using a theodolite, weather observers merely watch its ascent and plot its position. By assuming a standard rate of ascent, the direction and speed of the wind at various altitudes and the height of clouds can be determined. About 80 weather stations, and some ships at sea, launch these balloons (called *pibals* for short) at six-hour intervals. The disadvantage of the pibal is that it is usable only in daylight and good weather.

Radiosonde Observations

The radiosonde, first used in 1938, is a small, expendable, two-pound box containing temperature, humidity, and pressure measurement instruments and a miniature radio transmitter. This box is carried aloft by a large helium- or hydrogen-filled balloon, which also carries a parachute to lower the box safely to earth after the balloon has burst. The balloons rise to an average height of 15 to 20 miles, transmitting the three weather elements periodically via coded signals. Radiosondes provide some of the best and most reliable data that are obtained on upper-air conditions.

Rawinsonde Observations

Another excellent source of upper-air wind data is the rawinsonde. (Rawin is a contraction of "radar radio" and "wind.") Unlike the pibal, which cannot be used at night and whose use is sometimes restricted by cloud cover, a balloon-borne radiosonde can be tracked by a radar or radio direction finder to measure the direction and angular elevation of an ascending balloon, in addition to the data measured by a radiosonde, no matter what the conditions are, and to much greater ranges. If a radar tracking station is used, a radiosonde with a radar reflector is usually employed.

There are about 300 radiosonde and rawinsonde stations in the Northern Hemisphere. The rawinsondes provide information on winds, temperature, and pressure at very high levels, often above 100,000 ft.

These observations are usually made every 12 hours (0000 and 1200 GMT).

Weather Reconnaissance Flights

The U.S. Air Force, the U.S. Navy, and the U.S. Weather Bureau have special weather reconnaissance aircraft. These aircraft are specially equipped and specially manned, which in effect make them flying weather stations. They obtain weather and upper-air data in areas where such information is missing, or locate and track hurricanes and typhoons.

Meteorological Rockets

Meteorological rockets are operated by the Department of Defense and the National Aeronautics and Space Administration. They consist of a radiosonde fixed into the nose of a rocket. When the rocket reaches its maximum height, the nose cone ejects and frees the radiosonde, which also has its own parachute. As this instrument floats back to earth, it radios coded data to a mother station just as does the normal radiosonde.

Satellite Observations

The newest aid to the collection of weather data is the weather satellite which is capable of taking and transmitting photographs of cloud patterns at altitudes between 400 and 800 miles. Weather satellites also make other measurements, such as radiation, solar flares, etc., thereby increasing man's knowledge of the upper atmosphere.

The first weather satellite, *TIROS* I, was launched on April 1, 1960, and consisted of several instruments designed to make upper-air observations, plus cameras. It circled the earth for 78 days and transmited 22,952 cloud-cover pictures, an auspicious beginning.

In the last ten years, since only one-fourth of the earth has conventional weather reporting stations, weather satellites have proved invaluable in filling the gaps in the global weather picture. By analyzing cloud pictures over large areas, meteorologists are able to locate and predict the approach of dangerous storms which might otherwise go unreported.

While the first *TIROS* showed mostly cloud cover, the newer satellites are able to provide information on cloud type and thickness, wind direction, amount of water vapor, temperature, radiation, and electrical disturbances.

Thus, weather satellites are now providing increasingly sophisticated information—ice, snow, and cloud features, mountain waves, jet stream bands, squall lines, and, of course, outstanding pictures of hurricanes and tropical storms.

AMOS

Automatic Meteorological Observation Stations are becoming more numerous as equipment is improved and located. (The seagoing ones are called NOMAD, an acronym for Navy Oceanographic and Meteorological Device.) Automatic weather stations are now used at sea in such locations as the Gulf of Mexico, Norfolk, and San Diego, as well as land stations in the polar regions and other isolated areas. These stations measure air temperature, pressure, wind speed and direction, and ocean current every six hours. Some of these stations are powered by normal batteries; others by atomic batteries.

In addition to AMOS and NOMAD, other types of automatic equipment are used, and the reports from each type are slightly different. Many weather stations have automatic equipment which can measure runway visibility or runway visual range (if applicable), temperature and dew point, wind direction and speed, altimeter setting, and the amount of precipitation. Some equipment can even transmit reports automatically and provide printed copies.

Pilot Reports

One of the vital sources of weather information is pilots themselves. While ground observers at an airfield can measure ceiling, visibility, and those things they can see, the pilot is often best equipped to report on flight conditions aloft. After all, he is on the scene, or else he has just flown through the weather himself. Therefore, pilot reports (called *PIREPS*) of upper cloud height, icing, and turbulence, and other phenomena are of great value to forecasters, weather briefers, and other pilots. PIREP information is also valuable to meteorologists and traffic control personnel. PIREPS are made to the nearest Flight Service Station whenever unforecast weather or hazardous weather is encountered. Pilot reports are required to be solicited and collected by FAA stations under the following conditions:

(1) When thunderstorm activity, ceilings at or below 5000 ft, or visibilities at or below 5 miles are reported or forecast within a 100-mile radius of flight service stations.
(2) Hourly in offshore coastal areas, regardless of weather, by specified FSS as assigned by Regional Offices.
(3) When a Weather Bureau office of an FSS indicates a need because of specific weather in a flight assistance situation.
(4) When necessary to describe current flying conditions pertinent to natural hazards (mountain passes, ridges, peaks) between the station and the adjacent station.

Pilot reports of the following flight conditions are desired:

(1) If ridges and peaks are obscured, whether passes and valleys are open
(2) Cloud bases and tops, including trends
(3) Altitudes between layers where operation can be VFR
(4) Breaks in clouds where climb or descent VFR is possible
(5) Thunderstorm activity
(6) Icing
(7) Turbulence
(8) Information on safety of operation: strong winds, funnels, strong echoes on airborne radar, hail
(9) Temperature at cruising altitude or flight level and locations where marked changes occur

PIREPS are disseminated on the teletype services in two ways: (1) as PIREP summaries, and (2) as urgent individual PIREPS when they deal with hazardous weather. The latter ones are also included in the remarks section of aviation weather reports, and in the scheduled weather broadcasts of FSS. Those pertaining to tornadoes, hail, turbulence, icing, or other severe weather phenomena are broadcast immediately upon receipt.

A pilot making a PIREP should identify the location of the weather with respect to some well-known geographical point. Distances should be expressed in *nautical* miles, while visibility should be given in *statute* miles. The location of the weather should be followed by the time of the observation, the type of phenomena, the altitude of the phenomena in hundreds of feet above mean sea level, and, in reports of turbulence, icing, and condensation trails, the type of aircraft you are flying.

Here is how a typical PIREP might appear on the teletype:

DCA PIREP 1629 20 S DCA
1120E MDT RIME ICE 50 P2V

Here is the decode:

At 1629 GMT, a pilot flying a Navy P2V between
Richmond, Va., and Washington, D.C., reports
to Washington Radio that at 1620 GMT (1120 EST)
his aircraft encountered moderate rime ice at
5000 ft 20 miles south of Washington.

If you are departing from an airfield which carries PIREP information, take the time to read these reports as part of your weather briefing. And consider that the PIREP you make is a courtesy to other
pilots, one which may help him avoid some dangerous or unpleasant
weather.

AVIATION WEATHER OBSERVATIONS

Having covered the eleven most common *sources* of weather information, let us now review those elements of the weather which are of
particular interest and concern to the pilot. These are:

- Sky cover
- Ceiling
- Pressure and altimeter setting
- Wind (surface and upper air)
- Visibility
- Temperature and dew point

Sky Cover

The amount of sky cover is estimated by the weather observer in
tenths of the total sky, which is covered by clouds (either surface-
based or aloft) within the horizon. Four terms are commonly used to
describe sky cover: *clear, scattered, broken,* and *overcast.* If a sky is
less than one-tenth covered it is called clear; if it is one-tenth through
five-tenths covered, it is called scattered. From six-tenths to nine-
tenths covered, it is called broken, and above this, it is called overcast. (See Figure 12–9 for sky coverage symbols.)

Ceiling

Ceiling is defined as the height above the ground of the *lowest* layer
of clouds which, when taken in conjunction with the cloud layers

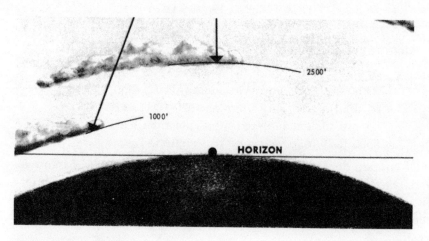

Figure 11–2 Definition of Ceiling.

below it, covers more than one-half of the sky (see Figure 11–2). When one-half or less of the sky is covered, the ceiling is said to be *unlimited*. If the cloud base cannot be seen at all either by heavy rain or fog, the ceiling is then said to be *zero*.

Ceilings are either "measured" or "estimated." Obviously, ones that are measured are more accurate than ones which are estimated.

There are several ways for measuring ceiling. One of the best ways is for the pilot who is climbing or descending to read the altimeter when the ground can be seen *straight down*. Another eyeball method in mountainous terrain is to estimate at what contour the mountains disappear. Experienced observers can usually make good ceiling estimates. A more accurate way is to release pilot balloons with a known ascension rate and observe them when they reach the cloud base. The time between release and entry into clouds provides a measurement of the cloud height. Radar on occasion can supply both the base and top of a cloud layer.

However, the best way of measuring cloud bases which is used at most stations is the ceilometer, which provides a continuous record of cloud heights and an accurate basis upon which to define ceilings.

In measuring the height of a cloud base with a ceilometer, a beam of high-power light is projected vertically against the cloud base. The height of the cloud base is automatically computed on the basis of (1) the known distance between the light projector and the observing point, and (2) the measured angle between the horizon and the place where the beam of light intersects the cloud. The important advantage of the ceilometer is that it may be used during daylight as well as at night.

The rotating-beam ceilometer scans continuously in a vertical plane. This ceilometer is usually located at the middle marker at most airports that are equipped with instrument landing systems.

Pressure and Altimeter Settings

The atmospheric pressure seen in an Aviation Weather Report is the pressure at the observing station reduced to mean sea level and reported in millibars and tenths. (Only three figures will be given on a weather report, so a pilot must decode it by placing a decimal to give tenths, and adding either "9" or "10" before the entry. To decide whether to use a 9 or a 10, remember that normal sea level pressure lies between 970 and 1060 mb. Thus, a reading of 137 would decode as 1013.7 mb.) Both aneroid and mercury barometers are used to measure the atmospheric pressure.

However, the pilot is more interested in the altimeter setting than he is in the pressure at sea level. Expressed in inches of mercury and reported to the nearest hundredth, the altimeter setting is simply the pressure measured at the station and reduced to a height of 10 ft above mean sea level. The altimeter indicates zero altitude at an elevation of 10 ft above mean sea level. Thus, at the height of 10 ft above the airport elevation (average cockpit height), the altimeter should indicate the airport elevation. The reduction of pressure read in the weather station to that at 10 ft above mean sea level is based upon the International Civil Aviation Organization Standard Atmosphere, which is the basis for altimeter design.

In the weather sequence report, the altimeter setting is the last item of the mandatory portion and immediately precedes any remarks given. To decode it, place a decimal between the last two figures and add either "2" or "3" in front. Remember that 28 to 31 in. is normal, so a number of 997 would become 29.97 in.

See Chapter 4 for a fuller discussion of this subject.

Wind

Wind direction, speed, character (gusty or squally), and wind shifts are determined by instruments at most weather stations. Wind vanes are used to measure direction and anemometers are used to measure speed. Surface wind direction is expressed by a two-digit number which converts to 36 sections of the compass. Thus, 21 would decode as 210°.

Wind direction is that direction from which the wind is blowing in reference to *true* north. Normally, it is the prevalent direction over a one-minute interval.

However, a pilot taking off or landing wants to know the wind direction in reference to *magnetic* north because runways are oriented on this basis. The wind direction stated in the regular space for wind information in Aviation Weather (Sequence) Reports is always in reference to *true* north, but the *magnetic* wind direction at the local airport is given in air-ground radio traffic between pilots and the tower or airport traffic control personnel. Wind directions at other locations are given in reference to true north.

Gustiness is defined as sudden, brief increases in wind speed. It is only reported when the variation between peaks and lulls is 9 knots or more. The average time interval between peaks and lulls usually does not exceed 20 seconds. When the wind is gusty, the speed of the peak gust during the past 15 minutes is reported following the usual report of average 1-minute speed. For example, "20G30" means that the 1-minute average wind speed is 20 knots, and the peak speeds in gusts are 30 knots.

(For a fuller discussion of the wind, see Chapter 6.)

Visibility

Of utmost importance to pilots is visibility—the horizontal distance at which mountains, TV and radio towers, and such obstructions to flight can be seen. Like ceiling, visibility can range between *zero* and *unlimited*. An unlimited visibility condition exists when one can see an object clearly at 10 miles or more. When the visibility is less than 7 miles, the sequence report will show certain symbols which show why the visibility is reduced (F = Fog, GF = Ground Fog, K = Smoke, etc.). These symbols are shown on page 203, Chapter 12.

Visibility is not always the same in all directions, however. Hence the term *prevailing visibility* is used in Sequence Reports. This is the greatest horizontal visibility in miles and fractions, which is equaled or surpassed throughout at least *half* of the horizon circle. The segments making up this half of the horizon circle need *not* be adjacent

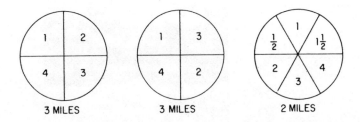

Figure 11–3 Determination of Prevailing Visibility.

to one another. Figures 11–3, 11–4, and 11–5 illustrate examples of the determination of prevailing visibility.

Runway Visibility

Runway visibility is the visibility observed along a specific runway expressed in miles and fractions. This is usually determined by a photo-electric device called a *transmissometer* which measures the transmissivity of the atmosphere along a known base length and extrapolates this measure to a visibility value along the runway. Runway visibility appears in the remarks section of Sequence Reports from authorized stations when the transmissometer value is less than two miles.

Another term a pilot will encounter is *runway visual range*. This is a runway distance derived by instruments and represents the horizontal distance a pilot can see down the runway from the approach end. It is based on the sighting of either high-intensity runway lights

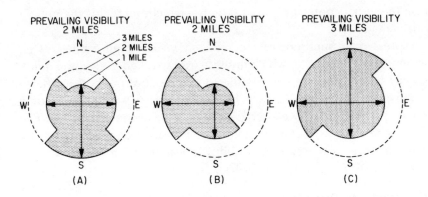

Figure 11–4 Determination of Prevailing Visibility. (The limits of visibility are represented by the shaded areas.)

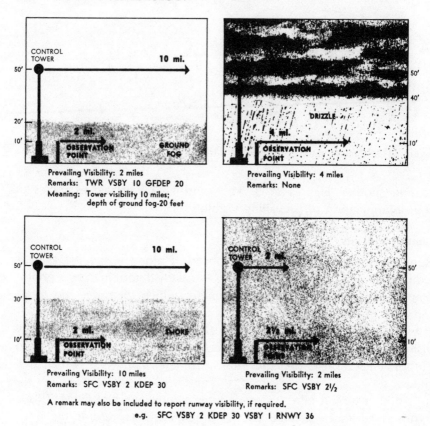

Prevailing Visibility: 2 miles
Remarks: TWR VSBY 10 GFDEP 20
Meaning: Tower visibility 10 miles;
         depth of ground fog-20 feet

Prevailing Visibility: 4 miles
Remarks: None

Prevailing Visibility: 10 miles
Remarks: SFC VSBY 2 KDEP 30

Prevailing Visibility: 2 miles
Remarks: SFC VSBY 2½

A remark may also be included to report runway visibility, if required.
e.g.   SFC VSBY 2 KDEP 30 VSBY 1 RNWY 36

Figure 11–5 Visibility at Different Levels.

or on the visual contrast of other targets—whichever yields the greater visual range.

Expressed in hundreds of feet, runway visual range capitalizes on the increased guidance which intense runway lights give the pilot. Runway visual range represents the most objective method in use to let the pilot know the distance he may expect to see when landing or taking off in poor weather conditions. It is reported as the first item in the remarks section of Sequence Reports from authorized stations whenever the prevailing visibility is less than 2 miles and/or the runway visual range is 6000 ft or less.

Temperature and Dew Point

Several types of recording sensors are used to measure temperature and dew point. The wet and dry bulb thermometer is the most com-

Figure 11-6 Hygrothermograph. This instrument records both the temperature and the relative humidity by tracing lines on a graduated chart fixed to a revolving drum. The temperature end is operated by the change in length of human hair, which is very sensitive to the moisture content of the atmosphere. (Courtesy Bendix-Friez.)

mon. Another, the hygrothermograph, is shown in Figure 11-6. The most advanced instrument, however, is the hygrothermometer which measures atmospheric moisture above relative humidities of 15%. Since it also provides temperature measurements, the hygrothermometer usually is installed near the center of the airport's runway complex in order to provide values representative of those actually experienced by aircraft on the runways. This is because the length of runway required for a jet aircraft takeoff is greatly influenced by the runway temperature and moisture content of the air.

Temperature and dew point are always reported in degrees Fahrenheit on the sequence report.

DISTRIBUTION OF WEATHER INFORMATION

The aviation weather information collected by the eleven sources just described is distributed by two principal means—teletypewriter and facsimile networks.

Figure 11-7 Service A Teletypewriter System. (Courtesy U.S. Weather Bureau.)

At every civilian weather information collection station, a weather observation (called the *ob*) is taken every hour. (Actually, most observers take them 10 minutes prior to the hour so the ob can be ready for transmission exactly *on* the hour.) After taking his readings, the observer cuts a paper tape on a special typewriter, inserts the tape into the teletypewriter and, exactly on the hour, initiates his transmission. At one of the five "interchange control centers," which are located at Atlanta, Cleveland, Kansas City, Fort Worth, and San Francisco, the individual weather station's call letters are activated, and the previously prepared weather report is typed on the station's own machine and instantaneously appears also on the many other weather stations on that particular circuit area. Figure 11–7 shows this teletypewriter Weather Bureau/FAA network.

Every Weather Bureau station within a given area, 8029 for example, receives reports from every other station. In addition, each station may receive reports from selected stations in other areas, the number received being dependent upon pilot briefing demands.

As you can see from Figure 11–7, there are 15 area circuits in the United States, each with as many as 120 weather reporting stations. The FAA operates these circuits, the Western Union or Bell Telephone Company owns and operates the equipment, while the Weather Bureau supplies the information which goes onto them.

In the United States, there are six teletypewriter networks, three civilian and three military.

The civilian networks are designated Service A, Service C and Service O.

Service A traffic consists of:

> Hourly and Special observations
> Pilot reports and summaries of pilot reports
> Winds aloft forecasts
> Radar summaries
> 12-hour terminal forecasts
> Aviation area forecasts
> Severe weather forecasts (WW)
> Inflight weather advisories (SIGMETS and ADVISORIES FOR LIGHT AIRCRAFT)
> Abbreviated hurricane advisories
> Regional weather synopses (Far West only)
> Simplified surface analyses and prognoses
> Notices to Airmen (NOTAMS)

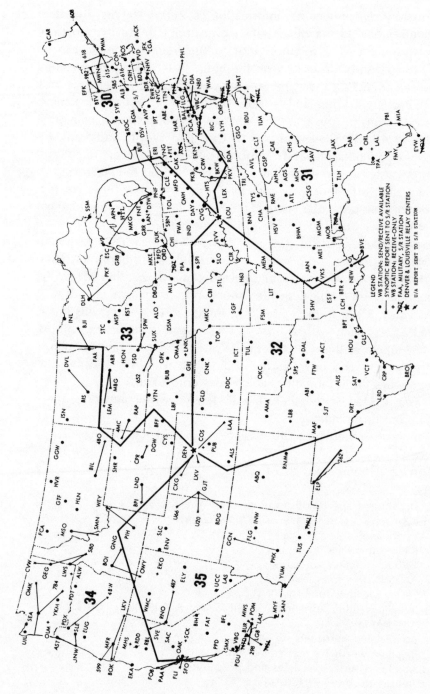

Figure 11-8 Service C Teletypewriter System. (Courtesy U.S. Weather Bureau.)

Service C is used as the primary national distribution system for aviation regional forecasts, 24-hour terminal forecasts, general service forecasts, upper-air observations, and nonaviation surface observations. The Service C network has six circuits shown in Figure 11–8.

Service O includes weather observations, coded analyses and prognoses, and forecasts, which are interchanged between the United States and foreign nations via the 6 domestic land-line and 13 international radioteletypewriter and cable teletypewriter circuits of the network. The domestic circuits extend to the approximately 100 locations that require overseas information.

Military Teletypewriter Networks

In the United States, the military services have their own teletype network, called *COMET*, which is similar to the civilian system described above. The U.S. Air Force system consists of three teletypewriter networks, each having eight circuits, as shown in Figure 11–9. (COMET

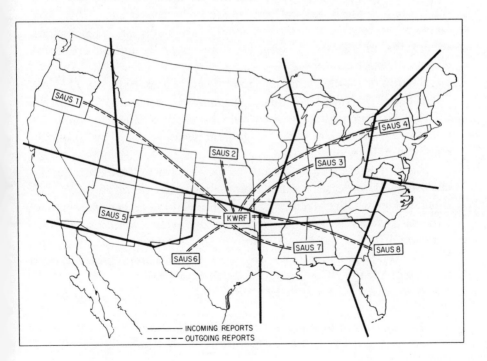

Figure 11–9 Military Teletype Area Map. (SAUS is an acronym for **S**urface **A**viation Observation **U**nited **S**tates.)

is an acronym from COntinental U.S. MEteorological Teletype systems.) COMET is used to collect pilot reports from military pilots and military weather observations in the contiguous United States and to provide rapid distribution of this information to military users. The three networks are controlled centrally from Carswell Air Force Base, and information is relayed to and from FAA circuits as needed. (All U.S. Naval Weather Service units in the United States file their observations on the COMET system.)

Weather Facsimile Networks

This is the second primary method of distributing aviation weather information. These networks distribute analyses and forecasts in graphic form and can also include special depictions of other types, such as photographs. *The National Weather Facsimile Network* is operated by the Weather Bureau and extends to about 360 civil and military weather offices in the contiguous United States (see Figure 11–10). The equivalent of 130 charts, measuring 12 in. by 18 in., is distributed by the National Weather Facsimile Network each day. The National Meteorological Center (NMC) originates all transmissions except for an average of nine per day made by the Severe Local Storms Forecast Center.

In addition to the national facsimile networks, both the U.S. Air Force and U.S. Navy have their own facsimile networks.

The Air Force Strategic Facsimile Network extends to some 80 military air bases and delivers graphic weather products oriented chiefly to military operations.

The six *Fleet Weather Centrals* of the U.S. Navy distribute weather analyses and forecasts for their respective areas of responsibility to smaller fleet weather facilities and to Navy ships at sea via separate all-weather radiofacsimile systems. These individual radiofacsimile systems are a part of the Navy's computerized worldwide combined all-weather facsimile and data network system. This computerized system links Fleet Weather Commands at Washington, D.C.; Monterey, California; Norfolk, Virginia; Pearl Harbor, Hawaii; Rota, Spain; London; and Guam.

The High-Altitude Facsimile Network is operated by the Weather Bureau and extends to 60 air terminals, with connections at some FAWS offices and military weather stations. Forecasts, distributed in graphic form, are for altitudes between approximately 18,000 and

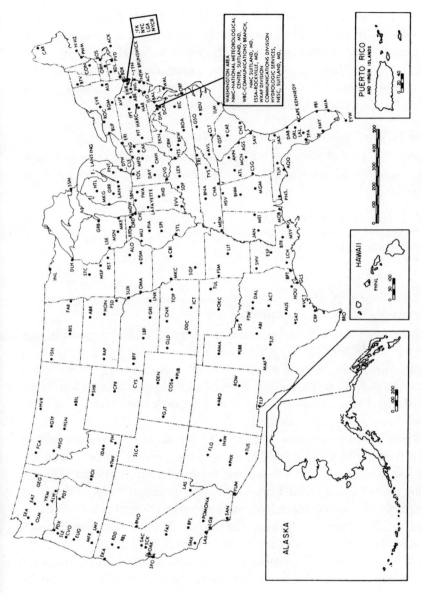

Figure 11-10 Weather Bureau Offices on National Facsimile Network.

45,000 ft. The products of all seven High-Altitude Forecast Centers placed together cover about two-thirds of the Northern Hemisphere. Long-line circuitry collects and distributes all material within the contiguous United States, and send-receive radiofacsimile systems provide links with foreign nations and Weather Bureau High-Altitude Forecast Centers at San Juan, Honolulu, and Anchorage.

Other Types of Aviation Weather Presentation

In addition to the products of the facsimile and teletypewriter networks described above, a pilot can also receive aviation weather by these means:

- Telephone answering service
- Person-to-person telephone briefing
- Recorded weather briefings
- Closed-circuit TV
- In-flight weather reports

The Pilot's Automatic Telephone Weather Answering Service (PATWAS) is a prerecorded briefing which is available to the pilot over his local telephone. In some areas, it is also available through a toll-free telephone service. (Such telephone numbers are not listed in telephone directories but are available in the *Airman's Information Manual.*) The military weather services also provide a recorded briefing of flight forecast information over a radius of 50 miles for military users through a similar method.

Pilots who fly only in a local area of certain large cities can obtain some weather assistance by dialing into the general-purpose weather forecast which is recorded by the local telephone company in 14 major cities from a script furnished by the Weather Bureau. Provided as a public service by the telephone company, this forecast is available at Baltimore, Boston, Chicago, Cleveland, Detroit, Los Angeles Milwaukee, New York City, Norfolk, Philadelphia, Pittsburgh, Richmond, San Francisco, and Washington. In a number of other cities, commercial sponsors provide a similar recorded general weather service by telephone.

Person-to-Person Telephone Briefing   The Weather Bureau provides weather briefings through both listed and unlisted telephone numbers. These are two-way conversations between the pilot and the briefer. The unlisted numbers for all Weather Bureau offices providing this service are published in the *Airman's Information Manual.* FAA Flight Service Stations also may be reached by telephone (numbers are listed in the *Airman's Information Manual* and in local telephone directories).

Pilot Briefing by Closed-Circuit Television   The Weather Bureau and FAA both have limited facilities for briefing by closed-circuit television at some large airfields. The pilot actually sees the charts on which the forecasts are made and hears them commented upon as well. Television briefing is second only to face-to-face briefing in effectiveness. The military weather services have been using this method of briefing effectively for some time.

In-Flight Weather   Weather information is received in flight by radio contact with Flight Service Stations of the Federal Aviation Agency. Locations of these facilities are shown in Figure 11–11. The Flight Service Specialist interprets the briefing material supplied by the

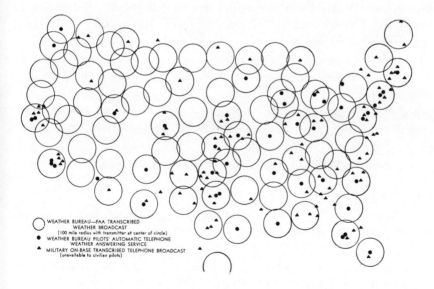

WEATHER BUREAU—FAA TRANSCRIBED
WEATHER BROADCAST
(100 mile radius with transmitter at center of circle)
WEATHER BUREAU PILOTS' AUTOMATIC TELEPHONE
WEATHER ANSWERING SERVICE
MILITARY ON-BASE TRANSCRIBED TELEPHONE BROADCAST
(unavailable to civilian pilots)

Figure 11–11   Transcribed Weather Broadcasts.

Weather Bureau when answering an in-flight request the same as he does when briefing over the telephone or face-to-face.

When filing a VFR flight plan, the pilot may request FAA's flight-following service. In addition to maintaining a radio watch for the pilot, the service relays pertinent weather information to him.

Transcribed Weather Broadcasts   Another in-flight service available on low- and medium-frequency navigation aids are the continuous transcribed weather broadcasts on Nav-Aid voice channels at 15 minutes past the hour. (In Canada, weather broadcasts start at 25 and 55 minutes past the hour.) These broadcasts contain a synopsis of current weather and forecasts of conditions for an area within a 250-mile radius of the station, as well as latest hourly reports from selected stations, pertinent pilot reports, and in-flight advisories (see below). The Weather Bureau provides the weather information, which is recorded by the FAA and re-broadcast on navigational aid frequencies. This service can be heard at distances from 150 to 250 miles.

In-Flight Weather Advisories   In-flight weather advisories serve the needs of airborne pilots and deal with weather expected in the next two to four hours. These advisories are of two types—the *SIGMET* (*Significant Meteorological Information*) and the *AIRMET* (*Airmen's Meteorological Information*). SigMets are applicable to all aircraft and concern hazardous weather such as tornadoes, squall lines, hail (¾ in. or more), severe and extreme turbulence, heavy icing, dust storms, and sandstorms.

SigMets are issued when any of the above conditions are considered imminent, preferably two hours in advance of the onset of any potentially hazardous weather.

A typical SigMet will appear on the teletypewriter like this:

FL JAX 152025
SIGMET ALFA 5. OVR ERN SC SRN GA JAX AREA AND ADJT
CSTL WATERS BNDD BY AREA ALG AND 50 MIS EITHER SIDE
OF A LN FM CROSS CITY FLA TO 30 N CHARLESTON SC FEW
SVR TSTMS AND ONE OR TWO TORNADOES HAIL TO 1 IN DIAM
ISOLD EXTRM TURBC SFC WNDS GUSTS TO 65 KT TIL 23Z.
ADNL AIRMET. ERN SC SRN GA JAX AREA AND ADJT CSTL

WATERS CIGS BLO 1 THSD FT AND VSBYS LCLY BLO 2 MIS
IN SCTD TO NMRS SHWRS AND TSTMS SOME IN SHORT LNS.
CONDS CONTG BYD 01Z.

FL .... Circuit heading identifying a SIGMET or AIRMET.
JAX .... ID of issuing FAWS
152025 .... Date-time group of transmission: 15th day
                 of the month at 2025Z

Here is the translation:

SIGMET ALFA 5—Numbered from midnight local standard
                 time
Over eastern South Carolina, southern Georgia, Jacksonville area, and
adjacent coastal waters bounded by area along and 50 miles either
side of a line from Cross City, Florida, to 30 miles north of Charleston,
South Carolina, few severe thunderstorms and one or two tornadoes,
hail to 1 in. in diameter, isolated extreme turbulence, surface winds
with gusts to 65 knots until 2300Z.

The in-flight advisory is broadcast by FSS at regular intervals on
the same navigation aid channels mentioned above.

AirMets are applicable mainly to aircraft having limited capability
because of lack of equipment or instrumentation, or pilot qualifica-
tion, and include weather phenomena such as moderate icing, moderate turbulence, extensive areas of visibilities less than 2 miles and
ceilings less than 1000 ft (including mountain ridges and passes), and
winds of 40 knots or more within 2000 ft of the surface.

A typical AirMet and its translation appear below.

FL SAT 190830
AIRMET DELTA 2. OVR SRN TEX E OF A LINE THRU MACALLEN-
AUSTIN CIG FQTLY BLO 1 THSD FT OCNL PATCHY FOG SERN
TEX VSBYS NEAR 2 MI

AIRMET DELTA 2—2nd advisory in the Delta series
for light aircraft issued by the San Antonio FAWS
since midnight local standard time on the 19th day
of the month.

Over southern Texas east of a line through Macallen-
Austin, ceiling frequently below 1000 ft.
Occasional patchy fog in southeastern Texas with
visibilities near 2 miles.

For the U.S. military pilot, many military bases provide a Pilot
Forecaster Service (PFSV). A special voice frequency is reserved for
this purpose and military pilots should make maximum use of it.

12 AVIATION WEATHER MAPS AND CHARTS

GENERAL

In the previous two chapters we have covered the aviation weather system and the recording, reporting, and distribution of aviation weather information. In this chapter, we will cover the various charts and maps which the pilot actually uses in planning and flying the weather.

SURFACE WEATHER MAPS

The primary weather aid which should be mastered by every pilot is the surface weather map. From it, he obtains an overall sea-level picture of the location of pressure systems and fronts. By referring to the last three or four surface weather maps, and by having an understanding of the physics of the atmosphere, a pilot can make his own mental forecast of the weather he can expect on his flight. Without

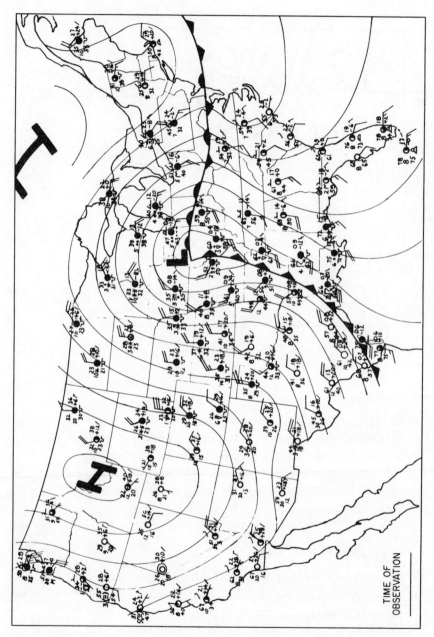

Figure 12-1 (A) Surface Weather Map.

TIME OF
OBSERVATION
_____

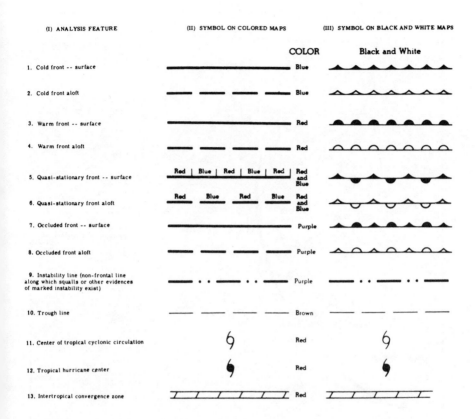

(I) ANALYSIS FEATURE   (II) SYMBOL ON COLORED MAPS   (III) SYMBOL ON BLACK AND WHITE MAPS

Figure 12–1 (B) Symbols Used on Surface Weather Maps.

a solid understanding of the surface weather map, he will not be fully capable of assimilating the other weather data (such as sequence reports, pireps, terminal forecasts) which are available.

There are two main types of surface maps. The colored "synoptic" surface map is plotted and hand-drawn by weathermen. This map is often "sectional," covering only a portion of the country (see Figures 12–1 and 12–2). The second type is the black and white (sometimes brown and white) facsimile map printed by a facsimile machine. Both types of these maps use the same source data from NMC. Since the colored map is larger, it is easier to read and, also, its station model data are more complete. Station models, and their differences, will be discussed below.

As described in Chapter 11, surface maps are the result of thousands of individual weather reports sent from a variety of sources—

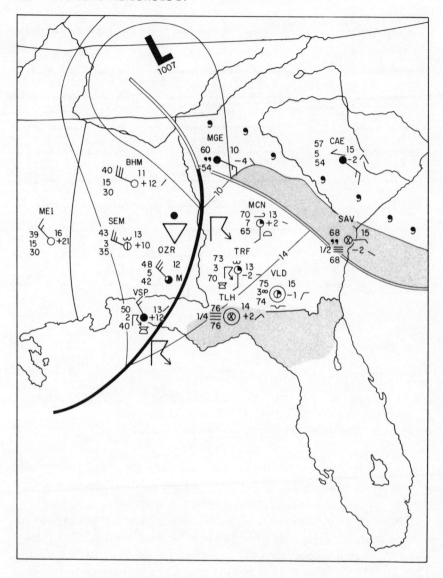

Figure 12–2   Sectional Surface Map.

approximately 350,000 per day. These reports are collected at the National Meteorological Center as follows: 900 airways surface reports every hour, 2000 international land station reports every 6 hours, 200 ship reports every 6 hours, 500 aircraft in-flight reports each day, 300 aircraft reconnaissance reports each day, 500 winds aloft reports every 6 hours, and 300 radiosonde reports every 12 hours. From these reports, surface maps are drawn and transmitted by facsimile machine. (These same data are also the source of the teletype sequence reports.) If a local weatherman does not have a facsimile receiver, or if his machine is broken, he uses the teletype sequence reports to draw his own surface weather map.

Pilots should know these things about surface weather maps:

- Since they are about two to six hours old by the time they are drawn, transmitted, and read by the pilot, they can be misleading. (Notwithstanding, these maps give the best and most complete weather picture over the entire United States)
- By comparing the latest surface weather map with earlier ones (most weather offices display the last three or four), a pilot can get an approximation of how the weather systems are progressing over or toward his planned route.
- Facsimile synoptic maps are drawn at the NMC eight times each day. Observations for these maps are taken at 0000Z, 0300Z, 0600Z, 0900Z, 1200Z, 1500Z, 1800Z, and 2100Z. This time appears in the lower left corner of the map.
- Remember these colors for weather and obstructions to visibility on a colored surface chart:

areas of precipitation	Green
(intermittent is green hatching; continuous is solid shading)	
fog, haze, and smoke	Yellow
dust or sand	Brown
severe weather or hazards to flying	Red

  (The complete color code is shown on Figure 12–1B.)
- Examine the isobar spacing. This will be a measure of the pressure wind gradient, and how fast the weather is moving.

Here are the things you should observe when you examine a surface weather map:

(1) Check the map time. How old is it?
(2) Note the position and movement trend of the lows which are along or near your planned flight route.
(3) Carefully examine the fronts. From an examination of previous surface maps, you can judge their rate and direction of movement and character.
(4) Locate your approximate route on the latest surface weather map. Estimate the position of any front or low-pressure system with regard to your planned flight route.
(5) Read the station models along both sides of your route. From these, you can get a good idea of the clouds you can expect, what ceilings

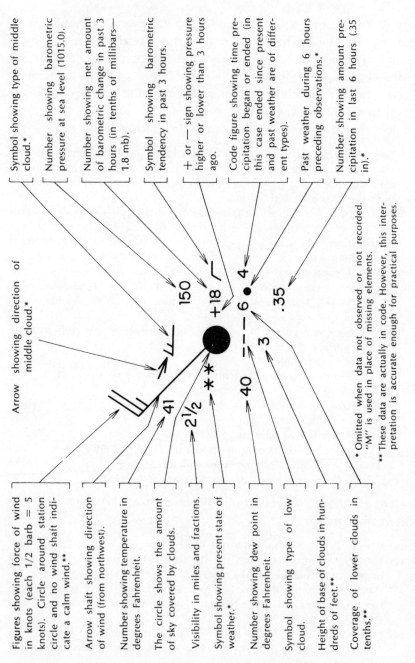

Symbol showing type of middle cloud.*

Number showing barometric pressure at sea level (1015.0).

Number showing net amount of barometric change in past 3 hours (in tenths of millibars—1.8 mb).

Symbol showing barometric tendency in past 3 hours.

+ or − sign showing pressure higher or lower than 3 hours ago.

Code figure showing time precipitation began or ended (in this case ended since present and past weather are of different types).

Past weather during 6 hours preceding observations.*

Number showing amount precipitation in last 6 hours (.35 in).*

Arrow showing direction of middle cloud.*

Figures showing force of wind in knots (each 1/2 barb = 5 knots). Circle around station circle and no wind shaft indicate a calm wind.**

Arrow shaft showing direction of wind (from northwest).

Number showing temperature in degrees Fahrenheit.

The circle shows the amount of sky covered by clouds.

Visibility in miles and fractions.

Symbol showing present state of weather.*

Number showing dew point in degrees Fahrenheit.

Symbol showing type of low cloud.

Height of base of clouds in hundreds of feet.**

Coverage of lower clouds in tenths.**

* Omitted when data not observed or not recorded. "M" is used in place of missing elements.

** These data are actually in code. However, this interpretation is accurate enough for practical purposes.

Figure 12-3 International Station Model.

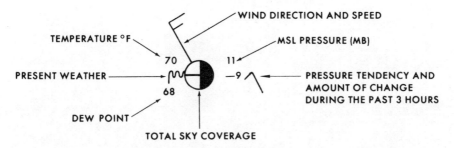

Figure 12–4 Facsimile Station Model.

you will encounter, whether thunderstorms, icing, fog, or turbulence
are present.
(6) Analyze the trends in temperatures and dew points by comparing the
last two maps. By doing so, a pilot can estimate the likelihood of
fog along his planned route.

THE STATION MODEL

To interpret a surface weather map, a pilot must understand the station model, which is the map's basic building block. It is a coded
picture of the existing weather at each individual surface station,
built around a circle, putting as much information as possible into a
small space. Around the circle is plotted other weather information
using figures and symbols.

Actually, there are three station models: (1) the international station model, (2) the facsimile station model, and (3) the sectional station model. Figure 12–3 shows a typical international station model,
while Figures 12–4 and 12–5 show the facsimile and the sectional sta-

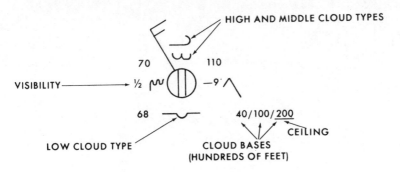

Figure 12–5 Sectional Station Model.

C_L LOW CLOUDS — Description (Abridged from I.M.O. Code)	C_M MIDDLE CLOUDS — Description (Abridged from I.M.O. Code)	C_H HIGH CLOUDS — Description (Abridged from I.M.O. Code)
Cu with little vertical development and seemingly flattened.	Thin As (entire cloud layer semitransparent).	Filaments of Ci, scattered and not increasing.
Cu of considerable development, generally towering, with or without other Cu or Sc bases all at same level.	Thick As, or Ns.	Dense Ci in patches or twisted sheaves, usually not increasing.
Cb with tops lacking clear-cut outlines, but distinctly not cirriform or anvil-shaped; with or without Cu, Sc, or St.	Thin Ac; cloud elements not changing much and at a single level.	Ci, often anvil-shaped, derived from or associated with Cb.
Sc formed by spreading out of Cu; Cu often present also.	Thin Ac in patches; cloud elements continually changing and/or occurring at more than one level.	Ci, often hook-shaped, gradually spreading over the sky and usually thickening as a whole.
Sc not formed by spreading out of Cu.	Thin Ac in bands or in a layer gradually spreading over sky and usually thickening as a whole.	Ci and Cs, often in converging bands, or Cs alone; the continuous layer gradually spreading over the sky, not reaching 45° altitude.
St or Sf or both, but not Fs of bad weather.	Ac formed by the spreading out of Cu.	Ci and Cs, often in converging bands, or Cs alone gradually spreading over the sky, the continuous layer exceeding 45° altitude.
Sf and/or Cf of bad weather (scud) usually under As and Ns.	Double-layered Ac or a thick layer of Ac, not increasing; or As and Ac both present at same or different levels.	Cs covering the entire sky.
Cu and Sc (not formed by spreading out of Cu) with bases at different levels.	Ac in the form of Cu-shaped tufts or Ac with turrets.	Cs not increasing and not covering entire sky; Ci and Cc may be present.
Cb having a clearly fibrous (cirriform) top, often anvil-shaped, with or without Cu, Sc, St, or scud.	Ac of a chaotic sky, usually at different levels; patches of dense Ci are usually present also.	Cc alone or Cc with some Ci or Cs, but the Cc being the main cirriform cloud present.

Cloud Abbreviation

St or Fs—Stratus or Fractostratus
Ci—Cirrus
Cs—Cirrostratus
Cc—Cirrocumulus
Ac—Altocumulus
As—Altostratus
Sc—Stratocumulus
Ns—Nimbostratus
Cu or Fc—Cumulus or Fractocumulus
Cb—Cumulonimbus

* Pilots should know these most common cloud symbols.

Figure 12-6 Cloud Symbols.

tion models. Basically, all three are alike, but the latter two contain less information.

Let us review a typical international station model, Figure 12–3.

Starting at the twelve o'clock position, the station model shows the cloud symbols for both the middle and high clouds. These symbols are shown one above the other, with the middle-cloud symbol closer to the station circle. (The 27 cloud types of the international code are shown in Figure 12–6. The most common ones which the pilot should know are marked by an asterisk.)

Proceeding clockwise around the station circle, there is a series of three digits at the one o'clock position. These digits express the station pressure in millibars. In order to read these three properly, place a decimal point between the second and third digit. If the first two digits of the three-digit group read any number between 00 and 50, place the number 10 in front of them. (For example: 150 would first read 15.0; then, since the first two digits fall between 00 and 50, place a 10 in front of the group and read 1015.0 mb.) If the first two digits read a number between 51 and 99, place a 9 in front of the group and read the corrected pressure. (For example, 792 would first read 79.2; then place a 9 in front, and read 979.2 mb of pressure.)

At the three o'clock position is a series of three symbols and numbers. This series gives the barometric change and tendency in the past three hours. The first of these three will be either a plus or minus sign, indicating a pressure rise or decline in the past three hours. The next part is generally composed of two numbers showing the amount

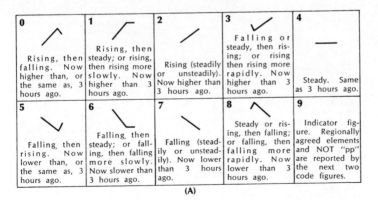

0	1	2	3	4
Rising, then falling. Now higher than, or the same as, 3 hours ago.	Rising, then steady; or rising, then rising more slowly. Now higher than 3 hours ago.	Rising (steadily or unsteadily). Now higher than 3 hours ago.	Falling or steady, then rising; or rising then rising more rapidly. Now higher than 3 hours ago.	Steady. Same as 3 hours ago.
5	**6**	**7**	**8**	**9**
Falling, then rising. Now lower than, or the same as, 3 hours ago.	Falling, then steady; or falling, then falling more slowly. Now slower than 3 hours ago.	Falling (steadily or unsteadily). Now lower than 3 hours ago.	Steady or rising, then falling; or falling, then falling more rapidly. Now lower than 3 hours ago.	Indicator figure. Regionally agreed elements and NOT "pp" are reported by the next two code figures.

(A)

Figure 12–7 (A) Barometric Tendency. (B) Present States of Weather Symbols. (C) Past States of Weather Symbols.

	0	1	2	3	4
00	Cloud development NOT observed or NOT observable during hour.	Clouds generally dissolving or becoming less developed during past hour.	State of sky on the whole unchanged during past hour.	Clouds generally forming or developing during past hour.	Visibility reduced by smoke.
10	Light fog.	Patches of shallow fog at station NOT deeper than 6 ft on land.	More or less continuous shallow fog at station, NOT deeper than 6 ft on land.	Lightning visible, no thunder heard.	Precipitation within sight, but NOT reaching the ground at station.
20	Drizzle (NOT freezing and NOT falling as showers) during past hour, but NOT at time of observation.	Rain (NOT freezing and NOT falling as showers) during past hour, but NOT at time of observation.	Snow (NOT falling as showers) during past hour, but NOT at time of observation.	Rain and snow (NOT falling as showers) during past hour, but NOT at time of observation.	Freezing drizzle or freezing rain (NOT falling as showers) during past hour, but NOT at time of observation.
30	Slight or moderate duststorm or sandstorm, has decreased during past hour.	Slight or moderate duststorm or sandstorm, no appreciable change during past hour.	Slight or moderate duststorm or sandstorm, has increased during past hour.	Severe duststorm or sandstorm has decreased during past hour.	Severe duststorm or sandstorm, no appreciable change during past hour.
40	Fog at distance at time of observation, but NOT ta station during past hour.	Fog in patches.	Fog, sky discernible, has become thinner during past hour.	Fog, sky NOT discernible, has become thinner during past hour.	Fog, sky discernible, no appreciable change during past hour.
50	Intermittent drizzle (NOT freezing) slight at time of observation.	Continuous drizzle (NOT freezing) slight at time of observation.	Intermittent drizzle (NOT freezing) moderate at time of ob.	Continuous drizzle (NOT freezing), moderate at time of observation.	Intermittent drizzle (NOT freezing), thick at time of observation.
60	Intermittent rain (NOT freezing), slight at time of observation.	Continuous rain (NOT freezing), slight at time of observation.	Intermittent rain (NOT freezing), moderate at time of ob.	Continuous rain (NOT freezing), moderate at time of observation.	Intermittent rain (NOT freezing), heavy at times of observation.
70	Intermittent fall of snowflakes, slight at time of observation.	Continuous fall of snowflakes, slight at time of observation.	Intermittent fall of snowflakes, moderate at time of observation.	Continuous fall of snowflakes, moderate at time of observation.	Intermittent fall of snowflakes, heavy at time of observation.
80	Slight rain shower(s).	Moderate or heavy rain shower(s).	Violent rain shower(s).	Slight shower(s) of rain and snow mixed.	Moderate or heavy shower(s) of rain and snow mixed.
90	Moderate or heavy shower(s) of hail, with or without rain, or rain and snow mixed, NOT associated with thunder.	Slight rain at time of observation; thunderstorm during past hour, but NOT at time of observation.	Moderate or heavy rain at time of ob.; thunderstorm during past hour, but NOT at time of observation.	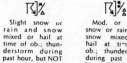 Slight snow or rain and snow mixed or hail at time of ob.; thunderstorm during past hour, but NOT at time of observation.	Mod. or heavy snow or rain and snow mixed or hail at time of ob.; thunderstorm during past hour, but NOT at time of observation.

Figure 12–7 (B)

5

Dry haze.

Precipitation within sight, reaching the ground, but distant from station.

Showers of rain during past hour, but NOT at time of observation.

Severe duststorm or sandstorm, has increased during past hour.

Fog, sky NOT discernible, no appreciable change during past hour.

Continuous rain (NOT freezing), heavy at time of observation.

Continuous rain (NOT freezing), heavy at time of observation.

Continuous fall of snowflakes, heavy at time of observation.

Slight snow shower(s).

Slight or mod. thunderstorm without hail, but with rain and/or snow at time of observation.

6

Widespread dust in suspension in the air, NOT raised by wind, at time of observation.

Precipitation within sight, reaching the ground, near to but NOT at station.

Showers of snow, or of rain and snow, during past hour, but NOT at time of observation.

Slight or moderate drifting snow, generally low.

Fog, sky discernible, has begun or become thicker during past hour.

Slight freezing drizzle.

Slight freezing rain.

Ice needles (with or without fog).

Moderate or heavy snow shower(s).

Slight or mod. thunderstorm, with hail at time of observation.

7

Dust or sand raised by wind, at time of observation.

Thunder heard, but no precipitation at the station.

Showers of hail, or of hail and rain, during past hour, but NOT at time of observation.

Heavy drifting snow, generally low.

Fog, sky NOT discernible, has begun or become thicker during past hour.

Slight freezing drizzle.

Moderate or heavy freezing rain.

Moderate or heavy freezing rain.

Granular snow (with or without fog).

Slight shower(s) of soft or small hail with or without rain or rain and snow mixed.

Heavy thunderstorm, without hail, but with rain and/or snow at time of observation.

8

Well developed dust devil(s) within past hour.

Squall(s) within sight during past hour.

Fog during past hour, but NOT at time of observation.

Slight or moderate drifting snow, generally high.

Fog, depositing rime, sky discernible.

Drizzle and rain, slight.

Rain or drizzle and snow, slight.

Isolated star-like snow crystals (with or without fog).

Moderate or heavy shower(s) of soft or small hail with or without rain or rain and snow mixed.

Thunderstorm combined with duststorm or sandstorm at time of observation.

9

Duststorm or sandstorm within sight of or at station during past hour.

Funnel cloud(s) within sight during past hour.

Thunderstorm (with or without precipitation) during past hour, but NOT at time of observation.

Heavy drifting snow, generally high.

Fog, depositing rime, sky NOT discernible.

Drizzle and rain, moderate or heavy.

Rain or drizzle and snow, mod. or heavy.

Ice pellets (sleet, U.S. definition).

Slight shower(s) of hail, with or without rain or rain and snow mixed, not associated with thunder.

Heavy thunderstorm with hail, at time of observation.

Sandstorm or duststorm.	Drifting or blowing snow.	Fog, or smoke, or thick dust haze.	Drizzle.
Rain.	Snow.	Shower(s).	Thunderstorm, with or without precipitation.

Figure 12–7 (C)

of barometric change in millibars and tenths in the past three hours (+18). The last part of this group shows pictorially the pressure tendency during the preceding three hours. In Figure 12–3, the barometric pressure has risen and become steady the last three hours. See Figure 12–7A for a list of these pressure tendency symbols.

On the outside at the five o'clock position is a group which shows past weather for the last six hours. This group contains a symbol indicating the type of past weather, a single digit indicating the time that precipitation began or ended, and the amount of precipitation expressed in hundredths of an inch (in this case, .35 in.). See Figure 12–7C for the complete list of past-weather symbols.

The six o'clock position has information relating to the type of low cloud and the amount of low-cloud coverage. It also indicates in hundreds of feet the height above the ground of the base of the lowest cloud in the sky (300 ft).

At the seven o'clock position is the dew-point temperature expressed in degrees Fahrenheit (40°). Above this, at the nine o'clock position, is found the present-weather symbol. Figure 12–7B is a list of the present state symbols.

Directly to the left of the present-weather symbol is the visibility, expressed in statute miles and fractions (2½). At the ten o'clock position is found the air temperature, expressed in degrees Fahrenheit (41°). (In other major countries, temperatures and dew points are in degrees centigrade.)

The wind information is indicated by the use of a wind arrow plus feathers or barbs. The arrow indicates the direction *from* which the wind is blowing and the barbs indicate wind speed. One full barb

◎	CALM
⌐	5 KNOTS
\	10 KNOTS
\	15 KNOTS
\	20 KNOTS
\	50 KNOTS

Figure 12–8 Station Model Wind Velocity Symbols.

indicates 10 knots, one-half barb 5 knots, and a pennant stands for 50 knots. A calm wind is indicated by a circle around the station model circle. Wind direction is plotted to the nearest 10° of the true compass and the speed is rounded to the nearest 5 knots. See Figure 12–8. (Chapter 6, page 77, includes more details of wind measurement and the Beaufort scale.)

The station circle itself is either partially or wholly filled in. This filling shows us the total amount of sky coverage produced by the existing clouds at the station. Figure 12–9 is the key.

There are many times when some of the information around a station model is omitted. Such an example would be the absence of middle- and high-cloud symbols if the low clouds completely covered the sky. The visibility is sometimes omitted when it is greater than 10 statute miles.

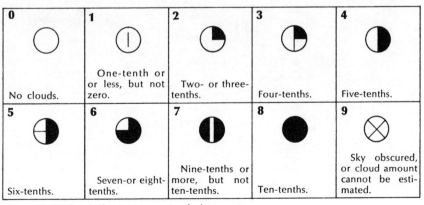

Figure 12–9 Total Sky Coverage Symbols.

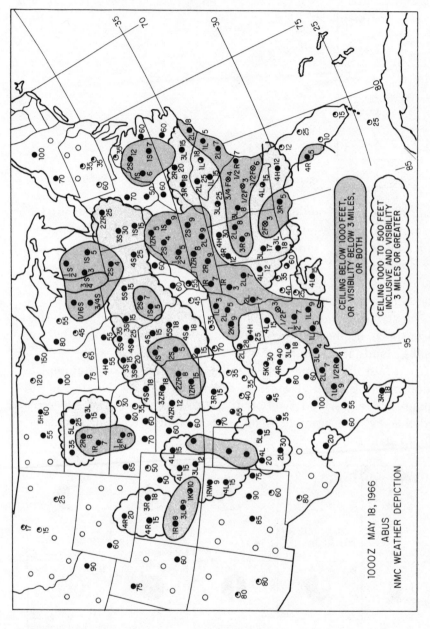

Figure 12–10 (A) Weather Depiction Chart.

1000Z MAY 18, 1966
ABUS
NMC WEATHER DEPICTION

CEILING BELOW 1000 FEET, OR VISIBILITY BELOW 3 MILES, OR BOTH

CEILING 1000 TO 500 FEET INCLUSIVE AND VISIBILITY 3 MILES OR GREATER

STATION MODEL	TOTAL SKY COVERAGE	SIGNIFICANT WEATHER
Significant weather — Total sky coverage / Vis to 6 mi — →3RF● / 15 ← Ceiling height (to 10,000)	◯ Clear / ◔ Scattered / ◑ Broken / ● Overcast / ⊗ Obscured	T Thunderstorm / ZR Freezing rain / ZL Freezing drizzle / R Rain / S Snow / E Sleet / L Drizzle / W Shower / H Haze / K Smoke / F Fog / GF Ground fog / ⟁ Clouds topping ridges

ANALYSIS:		
Ceiling and visibility classes:	Outline:	Color code:
(1) Ceiling below 1000 ft. or visibility below 3 miles, or both.	⬭	Red
(2) Ceiling 1000 to 5000 ft. inclusive and visibility 3 miles or greater.	⬭	Blue

NOTES:

(1) Precipitation will always be entered when reported as occurring at the station at the time of observation, except that no more than two symbols for significant weather will be entered.

(2) Obstructions to vision other than precipitation will be entered when visibility is reduced to 6 miles or less, except that if two forms of precipitation are reported, the other obstructions to vision will not be entered. Haze and smoke will be omitted if precipitation is reported.

(3) Visibility values of 6 miles or less will be entered in miles and fractions.

(4) All ceiling heights will be entered in hundreds of feet.

(5) The analysis of the ceiling-visibility classes will be based on the reports themselves and is not meant to include possible between-station orographic effects, or even systematic interpolations between stations.

Figure 12–10 (B) Symbolic Notation Used on Weather Depiction Charts.

WEATHER DEPICTION MAPS

One of the best displays of *current* weather (not forecast) is the Weather Bureau's Weather Depiction Chart (See Figure 12–10A.) From a glance at these "plan view" charts, a pilot can see areas where flights must be conducted under instrument flight conditions (less than 1000-ft ceiling, visibility less than 3 miles) and he can also see the marginal areas of possibly worsening conditions. In combination with the Surface Weather Chart and sequence reports, the Weather Depiction Chart gives the best combined picture of *present* weather that can be obtained. In other words, Weather Depiction Charts tell you *what*, surface maps tell you *why*. Data contained in these charts include:

- Precipitation and/or other important weather phenomena
- Sky coverage
- Visibility (when 6 miles or less)
- Cloud base (in hundreds of feet above the ground up to 20,000 ft)

The analysis consists of:

- *Solid* lines (called isolines) enclosing areas where ceilings are below 1000 ft and/or visibilities are below 3 miles. (Enclosed areas usually are shaded in red by the local weather station after they are received on the facsimile machine.)
- *Scalloped* isolines enclosing areas where ceilings are below 5000 ft but not below 1000 ft and visibilities are greater than 3 miles. (These areas usually are shaded in blue locally after they are received on the facsimile machine.)

Weather Depiction Charts are facsimile charts prepared by the National Meteorological Center every three hours beginning with the sequence reports data at 0100 GMT. These charts are sent to the various weather offices over the facsimile circuit.

The Weather Depiction Charts just described should not be confused with another type of depiction chart used by the U.S. Air Force, U.S. Army, U.S. Navy, and ICAO called *Horizontal Weather Depiction Charts* (HWD). These charts are 12-hour *forecast* charts, not current weather charts, which are valid at mid-time for a given flight. HWD charts are usually prepared as individual flight forecasts.

SEQUENCE REPORTS

As described in Chapter 11, the most rapid method of collecting and relaying present-weather data over a wide area is by teletype circuits. These circuits are also useful for amending forecasts and checking weather trends. Along with the surface and weather depiction maps, the *sequence report* (also called airway weather report) is the most valuable weather aid a pilot can have. Without a working knowledge of sequence reports, a pilot cannot make a good analysis of current weather and maps. They are so important, in fact, that every pilot should understand how they are prepared and transmitted.

There are two normal types of teletype sequence reports—*hourly reports* and the *special reports* called SCAN (*S*ignificant *C*hanges *A*nd *N*otices to Airmen). The latter sequences are made when ceiling or visibility changes are noted, tornadoes are observed, thunderstorm build-up or disappearance occurs, precipitation begins or ends, wind shifts, or similar conditions develop which might affect safety or efficiency of operations.

Let us now examine a typical United States sequence report in some detail.

```
023 SA23212100 (Weather Bureau) or SAUS 4 KWRF 212100Z (Military)
DIA 15ΘM30⊕1/2VRW−F 152/68/60/3018G30/996/VR28 WSHFT
1548E/⊖65/⊕VⓂ RB05 DRK NW VSBY.1/4V3/4 → DIA 9/5 QAPES →9/1
```

SYMBOL	ITEM	INTERPRETATION	TRANSLATION
	GROUP I	023 SA23212100 (Weather Bureau Sequence Report)	
023	Service A circuit identification	The last 3 digits of a circuit number containing a total of 4 digits.	Service A circuit number 8023 (The primary use of this first number group is as a guide to the communicator for the distribution to be made of the material).
SA	Traffic designator	Used to distinguish the material from other weather traffic on the circuit.	Hourly aviation collection with the specific meaning of Aviation Sequence.
23	Service A circuit identification	The last 2 digits of a circuit number containing a total of 4 digits.	Service A circuit number 8023 (Repeated primarily to identify the circuit for the user of the reports).
21	Date	The day of the month.	21st day of the month.
2100	Time of observation	Time of observation on the 24-hour clock in Greenwich mean time. To convert to local time, subtract 5 hours for eastern standard time, 6 hours for central, 7 hours for mountain, and 8 hours for Pacific.	4 P.M., eastern standard time.
	GROUP I SAUS 4 KWRF 212100Z (Military)		
SAUS	Military network area identifier	—	Surface Aviation Observation, United States.
4	Military teletype area number	—	
KWRF	Military teletype relay center		
21	Date	Day of month.	21st day of the month.
2100Z	Time of observation	Same as time above.	Same as time above. ("Z" means GMT.)

SYMBOL	ITEM	INTERPRETATION	TRANSLATION
		GROUP II DIA. 15⊕M30⊕1/2VRW–F	
DIA	Station identification letters	A 3-letter group in the United States identifying the station from which the report is sent. An attempt is made to assign designators that suggest the name of the sending station. Some examples are Oklahoma City (OKC), Tulsa (TUL), Richmond (RIC), Baltimore (BAL), San Francisco (SFO). As you can see, most civil station identification letters are abbreviations or the initials of the station. The USAF uses letters of the adjacent city while Navy identifiers start with "N."	Dulles International Airport Washington, D.C.
15⊕M30⊕	Sky cover and ceiling	The figures represent the height above ground of the bases of the cloud layers in hundreds of feet. The heights refer to the sky-cover symbols following the figures. ○ = Clear: less than 0.1 sky cover. ⊕ = Scattered: 0.1 to 0.5 sky cover. ⊕ = Broken: 0.6 to 0.9 sky cover (constitutes a ceiling unless preceded by –). ⊕ = Overcast: more than 0.9 sky cover (constitutes a ceiling unless preceded by –). – = Thin (when prefixed to the above symbols). If you can see the sun or moon through the cloud cover, it is called thin. + = Sky is unusually dark.	Scattered clouds at 1500 ft; overcast ceiling measured at 3000 ft.

SYMBOL	ITEM	INTERPRETATION	TRANSLATION
		GROUP II DIA. 15⊙M30⊕1/2VRW–F (Cont'd)	
		–X = Partial obscuration: 0.1 to less than 1.0 sky hidden by precipitation or obstruction to vision (bases at surface)	
		X = Obscuration: 1.0 sky hidden by precipitation or obstruction to vision (bases at surface) (constitutes a ceiling)	
		The ceiling figure will always be preceded by one of the following letters:	
		E = estimated; M = measured; W = indefinite; B = balloon; A = reported by aircraft; R = radiosonde or radar; D = estimated height of cirriform clouds; U = cirriform ceiling height unknown. When cirriform clouds do not constitute a ceiling and the height is unknown, the slant bar (/) is used before the sky cover symbol.	
		If the ceiling is below 3000 ft and is variable, the ceiling symbol will be followed by the letter V, and in the remarks the range of height will be indicated.	
	Visibility	Figures represent miles and fractions of statute miles. Followed by "V" if less than 3 miles and variable. If the visibility is 6 miles or less, the reason is always given under "Precipitation" or "Obstruction to vision."	
1/2V			Visibility ½ mile, variable.
	Precipitation, thunderstorm, or tornado	Weather Symbols A = hAil L = drizzLe	
RW–			Light rain shower.

SYMBOL	ITEM	INTERPRETATION	TRANSLATION

GROUP II   DIA. 15⊕M30⊕1/2VRW–F (Cont'd)

AP = small hail	R = Rain
E = slEet	RW = Rain shoWers
EW = slEet shoWers	S = Snow
IC = Ice Crystals	SG = Snow Grains
SP = Snow Pellets	SW = Snow shoWers
T = Thunderstorm	ZL = freeZing drizzLe
ZR = freeZing Rain	

(Tornado, funnel cloud, and waterspout are always spelled out.)

Intensities are indicated thus:
With the exceptions of A, AP, and IC, for which no intensity symbols are used, and for T, for which no "–" is used ("T" indicates a thunderstorm of moderate intensity or less), intensities are denoted by the following symbols:

– – Very Light	(no symbol) Moderate
– Light	+ Heavy

GROUP II   DIA. 15⊕M30⊕1/2VRW–F (Cont'd)

Obstruction to Vision Symbols

D = Dust	BN = Blowing saNd
F = Fog	BS = Blowing Snow
GF = Ground Fog	BY = Blowing spraY
K = smoKe	H = Haze
BD = Blowing Dust	IF = Ice Fog

(Plus and minus signs are not used with obstructions to vision.)

Obstructions to vision

F

Fog.

SYMBOL	ITEM	INTERPRETATION	TRANSLATION
		GROUP III 152/68/60/3018G30/996/	
		Most of the items in this group are separated by diagonal lines (/).	
152/	Pressure	Stated in millibars and tenths of millibars (omitting initial number "9" or "10").	Pressure 1015.2 mb.
68/	Temperature	In degrees Fahrenheit reported to nearest whole degree. Most other countries report in centigrade.	Temperature 68°F. A minus (−) sign before the numbers indicates below zero.
60/	Dew point	In degrees Fahrenheit	Dew point 60°F.
30	Wind direction (direction from which the wind is blowing)	To the nearest 10° starting at true north and moving clockwise through east and west, with true north reported as 360°. 2 digits are transmitted omitting the final zero; that is, 01 is used to represent 10°; 12, 120°; 28, 280°; etc. A calm wind is reported as a direction of 00.	Wind blowing from 300°.
18	Wind speed	Indicated in knots to the nearest knot. It is an average speed for a period of time, usually one minute. If the speed is estimated, the letter E will follow the value. If the wind is calm, the speed will be reported as 00.	Wind speed 18 knots.
G30/	Gustiness	Peak gust speed observed during the last 15 minutes prior to the time of observation. The letter Q in place of the G would indicate that squalls are present, followed by the peak speed in squalls during the same time interval used for gusts.	Wind gusts to 30 knots.
996/	Altimeter setting	Atmospheric pressure in inches and hundredths of inches of mercury for the setting of airin.	Altimeter should be set to 29.96 in.

SYMBOL	ITEM	INTERPRETATION	TRANSLATION
	GROUP III 152/68/60/3018G30/996/ (Cont'd)	craft altimeters. Given in three figures, omitting the initial number "2" or "3". A number beginning with 5 or higher presupposes a 2; a number beginning with 4 or lower presupposes a 3 (993 = 29.93 in., 002 = 30.02 in., etc.).	
	GROUP IV VR28 WSHFT 1548E/⊕65/⊕V⊕RBO5 DRK NW VSBY 1/4V3/4	This is the *remarks* section. Individual items are separated by diagonal lines (/) when they otherwise may be confused with one another.	
VR28	Remarks	The visual range down the instrument runway in hundreds of feet. For airports with multiple instrument runways and visual range measuring systems, VR22 would be a representative value obtained by averaging the visual ranges of all systems. If the visual range varies considerably from one runway to another, it is indicated separately with each runway identified. For example, "VR30 R18VR20 R10VR40" would be interpreted as "representative (average) visual range 3000 ft; runway 18, visual range 2000 ft; runway 10, visual range 4000 ft."	Runway visual range 2800 ft.
WSHFT 1548E/	Remarks	WSHFT is used to indicate that the wind direction has shifted; it is followed by the local standard time on the 24-hour clock that the wind shift occurred.	The wind direction shifted at 3:48 P.M., eastern standard time.

SYMBOL	ITEM	INTERPRETATION	TRANSLATION
	GROUP IV VR28 WSHFT 1548E/⊕65/⊕V⊕RBO5 DRK NW VSBY 1/4V3/4 (Cont'd)		
⊕65/	Remarks	Heights (above mean sea level) in hundreds of feet of bases and/or tops of sky-cover layers not visible at the station.	Top of overcast is 6500 ft.
⊕V⊕	Remarks	Remarks pertaining to coded elements reported in a preceding section.	The cloud layer based at 3000 feet with tops at 6500 ft is variable from overcast to broken.
RBO5	Remarks	Another remark relating to a coded element reported in a preceding section. This particular remark is in reference to the rain shower, meaning rain began at five minutes past the hour preceding the time of this observation. If rain had been reported on the last hourly report, but was no longer falling at the time of this observation, the remarks section would include the time that the rain (or other type of weather) ended by giving the letters RE, followed by the number of minutes after the last hourly report that the rain ended.	Rain began at 3:05 P.M., eastern standard time.
DRK NW	Remarks	An additional remark relative to the total weather picture in contractions of English words.	Dark to the northwest.
VSBY 1/4V3/4	Remarks	A remark relating to the prevailing visibility reported in Group I, giving the range in which the visibility is variable.	Visibility variable between 1/4 and 3/4 of a mile.

SYMBOL	ITEM	INTERPRETATION	TRANSLATION
		GROUP V → DIA 9/5 QAPES ↘ 9/1	
		When included, this section contains Notice to Airmen Information (NOTAMS) and is appended to the Aviation Weather Report following the last remark. Instructions for interpreting NOTAMS is usually available at weather briefing offices and may be obtained through reference to the FAA Air Traffic Service Handbook, Communications Procedures, AT P 7300.1.	
↑	NOTAM group	Formerly an arrow denoting a wind from the west, this arrow is now used in connection with the automatic processing of NOTAMS and has no significance to the pilot.	
DIA	NOTAM group	Repeat of the station identifier given at the beginning of the Aviation Weather Report.	The Notice to Airmen Information which follows applies to Dulles International Airport, Washington, D.C.
9/5	NOTAM group	The first digit designates the month of the year. The digit following the "/" indicates the NOTAM number. NOTAMS are numbered serially for the month, beginning with "1" for the first NOTAM. No arrow preceding this item of the NOTAM group indicates that is the first time that this NOTAM has appeared on the circuit.	This is the 5th NOTAM during the month of September which has been issued for Dulles International Airport.
QAPES	NOTAM group	This is the substance of the NOTAM, using the "Q" codes which have been employed for many years.	VOR and asociated voice communications out of service.

GROUP V → DIA 9/5 QAPES ↗ 9/1 (Cont'd)

SYMBOL	ITEM	INTERPRETATION	TRANSLATION
↗	NOTAM group	Current NOTAM indicator preceding the serial number of a previously transmitted NOTAM to indicate that it is still current.	Following is the serial number of a NOTAM for DIA which is still valid, but was transmitted earlier, and its contents will not be repeated in this report.
9/1	NOTAM group	These numbers are interpreted in the same manner as those preceding QAPES in this example.	The first NOTAM issued for DIA in the 9th month is still current. (The absence of 9/2, 9/3, and 9/4 indicates that those NOTAMS have been canceled.)

The following is a list of abbreviations which will be seen on the U.S. sequence reports:

List of Abbreviations

WORDS	CONTRACTIONS	WORDS	CONTRACTIONS
About	ABT	Condition	CND
Above	ABV	Continuing	CONTG
Accompany	ACPY	Correction	COR
Across	ACRS	Cumulonimbus	CB
Active	ACTV	Cumulus	CU
Advance	ADVN		
Advisory	ADVY	Daybreak	DABRK
After	AFT	Daylight	DALGT
After dark	AFDK	Decrease	DCR
Afternoon	AFTN	Decreasing	DCRG
Aircraft	ACFT	Deepen	DPN
Airline Reports	ALREPS	Delayed weather	PDW
Airmass	AMS	Dense	DNS
Airport	ARPT	Develop	DVLP
Airway	AWY	Dewpoint	DWPNT
Aloft	ALF	Diminish	DMSH
Along	ALG	Dissipate	DSIPT
Altimeter setting	ALSTG	Dissipating	DSIPTG
Amount	AMT	Distant	DSNT
Around	ARND	Divide	DVD
		Drift	DRFT
Bank	BNK	Drizzle	DRZL
Barometer	BRM	During	DURG
Become	BCM		
Becoming	BCMG	Early	ERY
Before	BFR	East	E
Begin	BGN	Ending	ENDG
Behind	BHND	Entire	ENTR
Below	BLO	Estimate	EST
Between	BTN	Evening	EVE
Break	BRK	Extend	XTND
Breaks in overcast	BINOVC	Extreme	XTRM
Broken	BRKN		
		Falling	FLG
Ceiling	CIG	Field	FLD
Center	CNTR	Flurry	FLRY
Central	CNTRL	Follow	FLW
Change	CHG	Forecast	FCST
Clear	CLR	Forenoon	FORNN
Clearing	CLRG	Forming	FRMG
Cloud	CLD	Freeze	FRZ
Coastal	CSTL	Frequent	FQT
Commence	CMNC	From	FM

List of Abbreviations (Continued)

WORDS	CONTRACTIONS	WORDS	CONTRACTIONS
Front	FNT	Lightning	LTNG
Frontal passage	FROPA	Likely	LKLY
Frost	FRST	Little	LTL
		Little change	LTLCG
Generally	GENLY	Local	LCL
Gradual	GRDL	Locally	LCLY
Ground	GND	Longitude	LONG
Ground fog	GNDFG	Lower	LWR
Group	GRP	Lower broken	LWRBRKN
		Lower overcast	LWROVC
Hailstone	HLSTO	Lower scattered	LWSCTD
Hard Freeze	HDFRZ		
Hazy	HZY		
Heavier	HVYR	Maximum	MAX
Heavy	HVY	Mean sea level	MSL
High	HI	Middle	MID
High broken	HBRKN	Midnight	MIDN
Higher	HIER	Mild	MLD
High overcast	HOVC	Mile (statute)	MI
High scattered	HSCTD	Miles per hour	MPH
Horizon	HRZN	Millibars	MB
Hundred	HND	Missing	MISG
Hurricane	HURCN	Mixed	MXD
Hurricane report	HUREP	Moderate	MDT
		Morning	MRNG
		Mostly	MSTLY
Ice on runways	IR	Mountain	MTN
Icing	ICG	Move	MOV
Icing in clouds	ICGIC		
Icing in precipita-		Nautical miles	NLM
tion	ICGIP	Night	NGT
Improve	IPV	North	N
Increase	INCR	Northern	NRN
Indefinite	INDFT	Numerous	NMRS
Inoperative	INOPV		
Instrument	INST		
Intense	INTS	Obscure	OBSC
Intermittent	INTMT	Observe	OBS
In vicinity of	INVOF	Occasion	OCN
		Occasional	OCNL
Knollsman	KOL	Occasionally	OCNLY
Knots	KT	Occluded front	OCFNT
		Occlusion	OCLN
Latitude	LAT	On top	OTP
Layer	LYR	Outlook	OTLK
Level	LVL	Over	OVR
Lift	LFT	Overcast	OVC
Light	LGT	Overhead	OVHD

List of Abbreviations (Continued)

WORDS	CONTRACTIONS	WORDS	CONTRACTIONS
Partly	PTLY	Snow	SNW
Persisting	PRSTG	Somewhat	SMWHAT
Period	PRD	South	S
Pilot balloon		Southeast	SE
observation	PIBAL	Spreading	SPRDG
Pilot reports	PIREPS	Sprinkle	SPKL
Portion	PTN	Squall	SQAL
Possible	PSBL	Squall lines	SQLNS
Precipitation	PCPN	Station	STN
Pressure	PRES	Storm	STM
Pressure rising		Stratus	ST
rapidly	PRESRR	Strong	SFC
Prevail	PVL	Surface	STG
Prognosis	PROG	Sunrise	SUNRS
Quadrant	QUAD	Temperature	TMP
Quadrants	QUADS	Tendency	TNDCY
		Terminal	TRML
Radar winds aloft	RAWIN	Thick	THK
Radiosonde	RA	Thin	THN
Radiosonde data	RADAT	Thousand	THSD
Ragged	RGD	Threaten	THTN
Range	RNG	Throughout	THRUT
Rapid	RPD	Thunder	THDR
Rapidly	RPDLY	Thunderhead	THD
Reach	RCH	Thundershower	TSHWR
Redeveloping	REDVLPG	Thunderstorm	TSTM
Reforming	REFRMG	Today	TDA
Region	RGN	Tonight	TNGT
Remain	RMN	Top of overcast	TOVC
Remark	RMRK	Topping	TPG
Ridge	RDG	Toward	TWD
Rising	RSG	Towering	TWRG
River	RVR	Trough	TROF
Runway	RNWY	Turbulence	TURBC
Scattered	SCTD		
Schedule	SKED	Unknown	UNK
Sections	SXNS	Unlimited	UNL
Several	SVRL	Unsteady	UNSTDY
Severe	SVR	Until	TIL
Shallow	SHLW	Upper	UPR
Shift	SHFT	Upward	UPWD
Shower	SHWR		
Sleet	SLT	Valley	VLY
Slightly	SLGTLY	Variable	VRBL
Smoke	SMK	Vicinity	VCNTY

List of Abbreviations (Continued)

WORDS	CONTRACTIONS	WORDS	CONTRACTIONS
Visibility	VSBY	West	W
Visible	VSB	Westerly	WLY
		Western	WRN
Warm	WRM	Westward	WWD
Weak	WK	Widely	WDLY
Weaken	WKN	Will	WL
Weather	WX	Wind	WND

WORD ENDINGS

Able	BL	lest, est	ST
Al	L	lness, ness	NS
Ally, erly, ly	LY	Ing	G
Ance, ence	NC	Ity	TY
Ary, ery, ory	RY	Ive	V
Der	DR	Ment	MT
Ed, ied	D	Our	US
Ening	NG	S, es, ies	S
Er, ier, or	R	Tion, ation	WD
Ern	RN	Ward	N
Ically	CLY		

Shown below are four United Kingdom Aviation Sequence Reports, which are known as *METARs*.

SAUK EGLL 220900Z
METAR
EGLL 31004 0200 R0300/27 45FG 9/00 13/11 1024
EGPH 24001 0600 R0790 50DZ 8ST03 12/11 1022/3017INS
EGGP 08024/40 8000 2SC50 6AC60 01/MO5 1021
EGCC 13018 1500 R2000/07 72SN 2ST30 5NS63 999

The decodes follow.

EGLL	ICAO identifier for London (Heathrow)
31004	Wind from 310°T at 4 knots
0200	Horizontal surface visibility, 200 meters
R 0300/27	Runway Visual Range (RVR), 300 meters on runway 27
45 FG	Present weather, fog (note: either code numbers, letters, or both, can be used at the option of the national weather service)
9//00	Sky obscured, cloud types unknown, height lowest cloud layer less than 30 meters
13/11	Air temperature, 13°C.; Dew-point temperature, 11°C.
1024	QNH 1024 mb

EGPH	ICAO identifier for Edinburgh (Turnhouse)
24001	Wind from 240°T at 1 knot
0600	Horizontal surface visibility, 600 meters
R 0790	Runway Visual Range (RVR), 790 meters (runway unspecified)
50 DZ	Present weather, drizzle
8 ST 03	8-eighths sky cover of stratus type at 90 meters (approximately 300 ft)
12/11	Air temperature, 12°C.; Dew-point temperature 11°C.
1022/	
3017INS	QNH 1022 mb or 30.17 in.

EGGP	ICAO identifier for Liverpool
08024/40	Wind from 080°T at 24 knots, maximum wind 40 knots
8000	Horizontal surface visibility, 8000 meters
——	RVR not reported or not installed
——	No significant present weather
2 SC 50	2-eighths stratocumulus at 1500 meters (approx. 5000 ft)
6 AC 60	6-eighths altocumulus at 3000 meters (approx. 10,000 ft)
01/MO5	Air temperature 1°C.; Dew-point temperature −5°C.
1021	QNH, 1021 mb

EGCC	ICAO identifier for Manchester
130 18	Wind from 130°T at 18 knots
1500	Horizontal surface visibility, 1500 meters
R 2000/07	Runway Visual Range (RVR), 2000 meters on runway 07
72 SN	Present weather, snow
2 ST 30	2-eighths stratus at 900 meters
5 NS 63	5-eighths nimbostratus at 3900 meters
999	QNH, 999 mb

Below is a typical Canadian Sequence Report and its decode:

DH 5⊕E 15⊕12RW- 1∅4/37/35/3616/982/SF1 SC9 CLD ON HILLS E
OCNL SW-

DH	Daniels Harbor, Newfoundland
5①	Scattered clouds at 500 ft above terrain
E	Estimated
15⊕	Overcast cloud layer at 1500 ft above terrain
12	Visibility 12 miles
RW-	Light rain showers
1∅4	Sea-level pressure 1010.4 mb
37	Temperature 37°
35	Dew point 35°
3616	Wind direction 360° at 16 knots
982	Altimeter setting 29.82 in.
SF1	Stratus fractus 1/10 (reference to 50)
SC9	Stratocumulus 9/10 (reference to 150)

CLD ON
HILLS E   Clouds on the hills east of the station
OCNL   Occasional
SW-   Light snow showers

Using Sequence Reports

As stated earlier, when a pilot goes into a weather office, he should review the current-weather maps first, starting with the latest and then examining two or three older ones. This will give him an overall view of the *current* weather in his proposed area of flight and how it is trending.

Some pilots, however, assume that surface maps are "too old," and tend to depend only on sequence reports, thinking them a faster means of "getting the picture." This can be a dangerous practice, however, since it can be very difficult to establish trends solely by sequence reporting. Moreover, by depending solely on sequence reports for flight preparation, and making only scanty reference to current maps and charts, a pilot will have to study many more sequence reports, not only along his proposed flight track, but for many miles on each side of it. On the other hand, time will be saved and accuracy gained by first studying the latest weather maps, and then checking the sequence reports.

Read the sequence reports for your point of departure, several en route weather stations, including some to each side of track, the destination, and, of course, the alternate.

RADAR SUMMARY CHARTS

The fourth type of current (not forecast) weather chart is the Radar Summary Chart. These charts are depictions of severe weather determined by radar from selected radar sites across the United States. Hourly reports (usually from stations east of the Rockies) are collected and analyzed at Kansas City by the Radar Analysis and Development Unit. A Radar Summary Chart and the symbols used thereon are shown in Figure 12–11.

The pilot should note these points about Radar Summary Charts:

(1) Check the time of the radar observations which appear in the lower left area of the chart.
(2) Scalloped lines indicate areas of heavier radar returns.

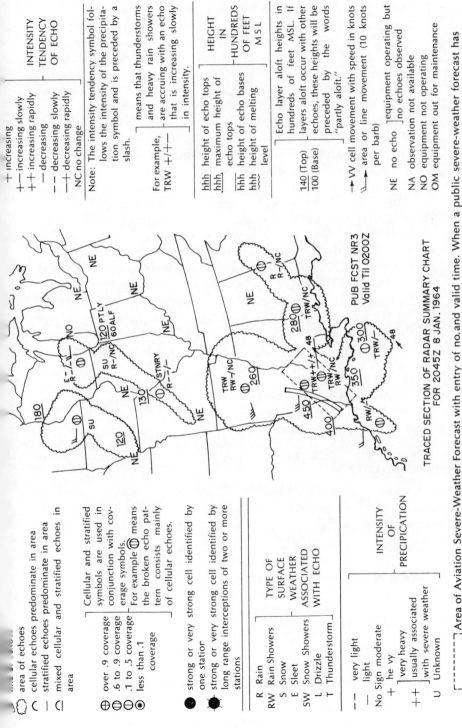

+ increasing
+- increasing slowly
++ increasing rapidly
− decreasing
−− decreasing slowly
−+ decreasing rapidly
NC no change
} INTENSITY TENDENCY OF ECHO

Note: The intensity tendency symbol follows the intensity of the precipitation symbol and is preceded by a slash.

means that thunderstorms and heavy rain showers are accruing with an echo that is increasing slowly in intensity.

For example, TRW +/+—

hhh height of echo tops
hhh maximum height of echo tops
hhh height of echo bases
hhh height of melting level
} HEIGHT IN HUNDREDS OF FEET MSL

140 (Top)
100 (Base)
Echo layer aloft heights in hundreds of feet MSL. If layers aloft occur with other echoes, these heights will be preceded by the words "partly aloft."

→ VV cell movement with speed in knots
↘ area or line movement (10 knots per barb)

NE no echo] equipment operating but no echoes observed
NA observation not available
NO equipment not operating
OM equipment out for maintenance

area of echoes
C cellular echoes predominate in area
— stratified echoes predominate in area
C(mixed cellular and stratified echoes in area

Cellular and stratified symbols are used in conjunction with coverage symbols. For example ⊕ means the broken echo pattern consists mainly of cellular echoes.

⊕ over .9 coverage
⊖ .6 to .9 coverage
⊘ .1 to .5 coverage
⊙ less than .1 coverage

● strong or very strong cell identified by one station.
⊛ strong or very strong cell identified by long range interceptions of two or more stations

R Rain
RW Rain Showers
S Snow
E Sleet
SW Snow Showers
L Drizzle
T Thunderstorm
} TYPE OF SURFACE WEATHER ASSOCIATED WITH ECHO

−− very light
− light
No Sign moderate
+ heavy
++ very heavy
  usually associated with severe weather
U Unknown
} INTENSITY OF PRECIPITATION

PUB FCST NR3
Valid Til 0200Z

TRACED SECTION OF RADAR SUMMARY CHART
FOR 2045Z 8 JAN. 1964

[──────] Area of Aviation Severe-Weather Forecast with entry of no. and valid time. When a public severe-weather forecast has been issued the lines defining the area dashed.

Figure 12–11 Explanation of Radar Summary Chart.

(3) Unbroken, double-line areas represent squall lines.

(4) Scattered, broken, and overcast sky coverage symbols are used to indicate scattered, broken, and continuous radar returns. An inverted "U" above the symbol indicates that clouds are building upward.

(5) An underlined number indicates the maximum altitude of the radar return in hundreds of feet above sea level. If a number is found below the line, it will indicate the minimum altitude of the radar return.

(6) The arrow with a number indicates the *expected* direction of movement and speed of the echoes.

(7) The arrow with barbs indicates the *actual* movement of echoes. Half-barb for 5 knots, full barb for 10 knots, etc.

(8) A *dotted line* outlines severe weather warning (WW) areas. WW# identifies the severe weather warning message used as a basis for drawing the severe weather warning area on the chart. Valid time will be shown opposite the corresponding WW# near the lower left portion of the chart.

(9) Letter code:
  (a) NR—no report.
  (b) NE—no echoes.
  (c) NA—not available.
  (d) VW—very weak echoes.
  (e) W—weak echoes.
  (f) W⁻—weak echo weakening.
  (g) W+—weak echo increasing.
  (h) M—moderate echo.

FREEZING-LEVEL CHART

This chart is used to determine the altitude of the freezing level. It is prepared and transmitted every 12 hours. Figure 12–12 is an example. Note the following:

- The surface freezing line is represented by dashed lines.
- The freezing level at altitude is represented by solid lines drawn at each 4000 ft of altitude, and each line is labeled.
- The three-digit numbers on the chart denote the altitude of the freezing level above sea level at the reporting station in hundreds of feet. In Figure 12–12, for example,

023 is 2300 ft
052 is 5200 ft
104 is 10,400 ft

- The time of the map is shown in the lower left corner.

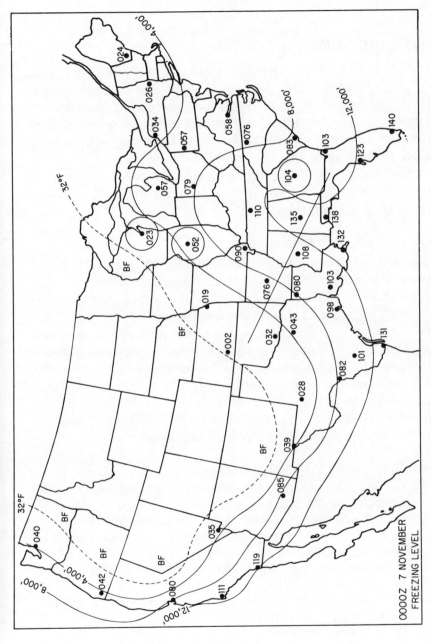

Figure 12–12 Freezing-level Chart.

WINDS ALOFT CHARTS

The wind at altitude is an important factor in flight and navigation. Every pilot should take full advantage of winds aloft information, both current and forecast, in order to (1) compute correct headings; (2) determine the best altitudes for flight; (3) compute range, ground-speed, and estimated time of arrival; and (4) be prepared for wind velocity changes along the flight route.

A pilot should also remember he cannot always be cleared at the "best wind" altitude. He should therefore write down the expected winds at all possible flight altitudes.

Collecting Winds Aloft Observations

Winds aloft are determined by pilot balloon (PIBAL) and rawinsonde observations as described in Chapter 11. These observations are taken at approximately 180 weather stations in the United States and are transmitted four times a day on the teletype circuits. They are transmitted in either one of two multi-part codes adaptable to both land and ship upper-wind observing stations. If in fact the observation is in the form of a rawinsonde, the data, including wind data, are coded in a complex form known as *TEMP*, which is better left to the professional analyst to decode. If the observation is a PIBAL or RAWIN only, it is coded in a form known as *PILOT* which most generally takes the symbolic form and is decoded as follows:

1st Section—QQ YYGGa$_4$ IIiii where:
(1) QQ is a message indicator specifying that wind data are for levels below 100,000 ft.
(2) YY is the date of the month on which the observation is taken to which 50 has been added to indicate that wind speeds are reported in knots.
(3) GG is the actual time of observation to the nearest whole hour in GMT.
(4) a$_4$ is the indicator specifying type of wind-measuring equipment (may be ignored).
(5) IIiii is a five-digit number assigned to the observing station.
2nd Section—9t$_n$u$_1$u$_2$u$_3$ ddfff ddfff ddfff where:
(1) 9 is the indicator that altitudes are in 1000-ft increments.

(2) t_m indicates the tens digit of altitude (expressed in 1000-ft increments which applies to data groups following.

(3) $u_1u_2u_3$ are units digits of the altitude (expressed in 1000-ft increments) applying to the first, second, and third data groups following, respectively.

(4) dd is the hundreds and tens digits of the true direction from which the wind is blowing after having been rounded to the nearest 5° (for example, 293 is rounded to 295 and coded 29; 297 is rounded to 300 and coded 30).

(5) fff is the direct-reading wind speed in knots when the rounded wind direction units digit is 0; when the rounded wind direction units digit is a 5, fff is coded as the wind speed plus 500 (for example, if the rounded wind direction 295° and the speed is 162 knots, fff is 162 + 500 = 662).

(6) The second section is repeated as many times as necessary to include all wind data.

Sample Winds Aloft Report (with corresponding symbolic form)

QQ   YYGGa₄   IIiii $9t_nu_1u_2u_3$ ddfff   ddfff
    65122   74222   90012   18030 17043

ddfff $9t_nu_1u_2u_3$ ddfff   ddfff   ddfff
16550   90345   16055   15070   13060

etc etc etc . . . $9t_nu_1u_2u_3$ ddfff
              92012   09525

ddfff   ddfff   etc etc . . .
08015   00000

Interpretation of Winds Aloft Reports

In the example shown, the upper-wind observation reports data below 100,000 ft (QQ) for the 15th day of the month (15 + 50 = 65, the addition of 50 indicating that wind speed is coded in knots). The data come from the 1200 GMT observation (12). The final coded value, 2, in this group can be ignored. The station is identified as 74222 (sometimes this is shortened to the iii form). The third five-digit group indicates (2) that data levels are in 1000-ft increments (9), (b) that the first following five-digit group is surface data (0 and 0), (c) that the second following five-digit group is the 1000-ft level (0 and 1), and (d) that the third following five-digit group is the 2000-ft level (0 and 2). These three data levels are then decoded as: 180° = 30 knots,

$170° = 43$ knots, and $165° = 50$ knots (note that fff was coded as $500 + 50$ indicating rounded wind direction was 160°), respectively. The seventh five-digit group (the 9 indicator is repeated) indicates that the following three levels are for the 3000, 4000, and 5000-ft levels, the data being $160° = 55$ knots, $150° = 070$ knots, and $130° = 060$ knots at these levels, respectively. Note that the group, 92012, indicates data follow for the 20,000, 21,000, and 22,000-ft levels. In this manner a wind report is coded for as many levels as data are available. A calm level is coded 00000. Missing data are indicated by the slant or solidus, /.

Although a certain facility and pride are developed in time with the ability to read the individual wind report, the reports are most valuable when plotted on an area chart and viewed as a group. Individual reports may then be better evaluated in space and time, and the state of the atmosphere and its changes can be determined more systematically. The pilot is encouraged to use the plotted winds aloft charts and forecasts in his flight briefing in preference to translating individual coded reports.

Winds Aloft Charts

The National Meteorological Center prepares winds aloft charts from the winds aloft reports. These charts are transmitted four times daily, approximately two hours after the standard PIBAL observation times. The observed winds for appropriate levels are plotted on these charts, using the barb and pennant system. The wind direction to the nearest 10° is represented by an arrow flowing with the wind and a numeral near the tail of the arrow. Figures 12–13, 12–14, and 12–15 show the three levels plotted.

Levels Plotted Winds aloft charts are transmitted over facsimile circuits within the United States. While there are three levels plotted— low, intermediate, and upper—the first two are of greatest interest to pilots.

The lower-level chart containing plotted winds for the second standard level above the surface (approximately 2000 ft above the ground), 5000, 8000, and 10,000-ft altitudes above mean sea level.

The intermediate-level chart containing winds for the 14,000, 20,000 25,000, and 30,000-ft altitudes above mean sea level.

The upper-wind-level charts for winds at 35,000, 40,000, 50,000, and the tropopause are used mainly for high-altitude commercial jets, military aircraft, and for research purposes.

Using the Charts Winds aloft charts are useful in representing the wind-flow pattern that existed at the time of the PIBAL or RAWIN observation. However, the charts are based on observed data several hours old and may not represent the wind that will be encountered on a flight subsequent to the observation time. For this reason, the pilot should also refer to the winds aloft *forecast* reports when planning his flight.

CONSTANT-PRESSURE CHARTS

As stated earlier, the Surface Weather Chart is a picture of current or past atmospheric conditions from the surface to approximately 5000 ft. Above 5000 ft, the pressure patterns and frontal positions begin to change shape and location as the surface secondary circulation of moving highs and lows gradually merges into the more constant circulation of the upper atmosphere. Constant-pressure charts (actually they are constant-pressure-*height* charts, and they are also known as pressure-contour charts) are weather maps designed for flights at and above 5000 ft. They depict the contoured flow along a specific pressure height level, in contrast to Surface Weather Maps which depict the pressure changes at a constant altitude. Furthermore, many weather phenomena occur at high altitudes which are unique and which do not appear on Surface Weather Maps, for example, the jet stream and clear air turbulence. Figure 12–16 is an example of a 500-mb chart.

In addition, there are times when widespread weather may hinder or prohibit the collection of winds aloft information. When adequate wind information is not available for winds aloft charts, constant-pressure charts must be used to determine wind conditions aloft over the oceans and in sparsely populated areas. In this case, they are often the only source of wind information.

When a pilot is planning a flight near the middle and upper portions of the troposphere, he should use the constant-pressure charts in addition to the Surface Weather Map. The constant-pressure charts correlate several weather facilities in the weather station and are drawn twice each day from observations made at 0000Z and 1200Z.

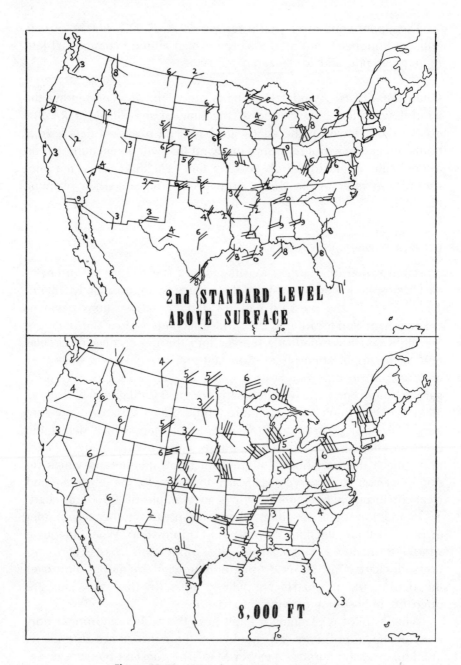

Figure 12–13 Lower-levels Winds Aloft Chart.

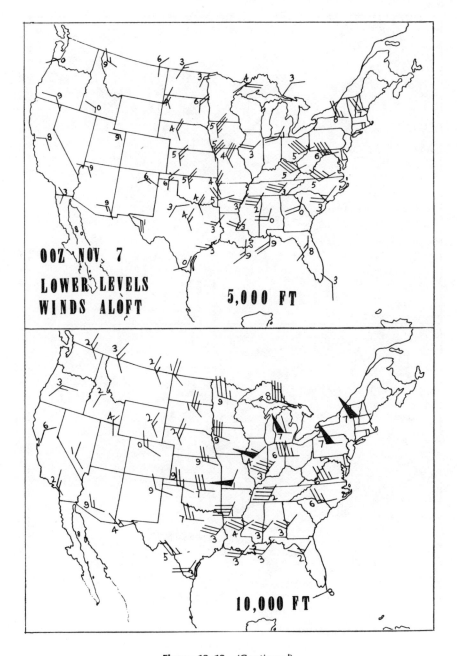

Figure 12–13   (Continued)

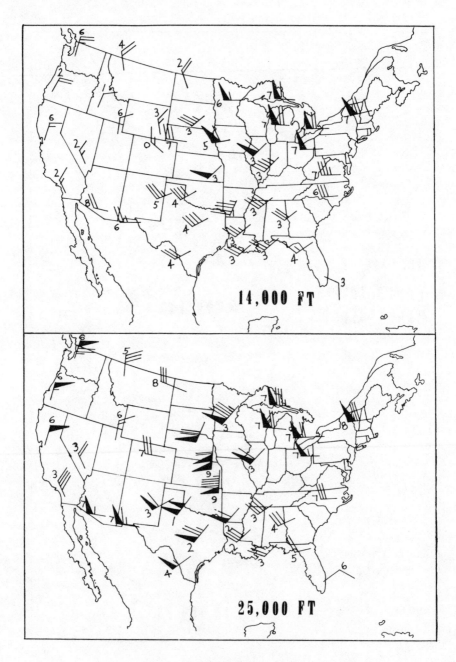

Figure 12–14 Intermediate-levels Winds Aloft Chart.

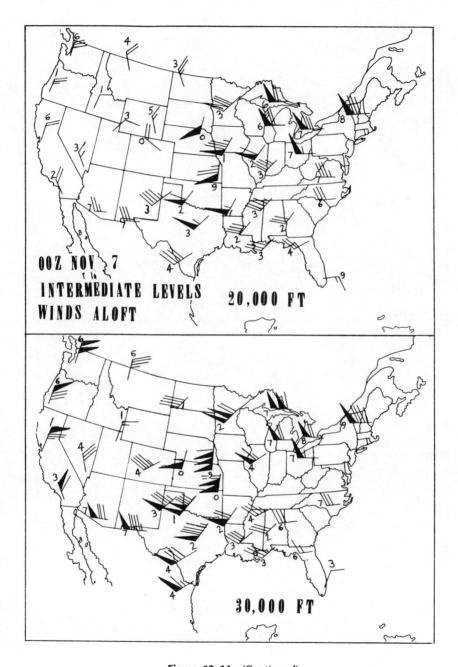

Figure 12–14 (Continued)

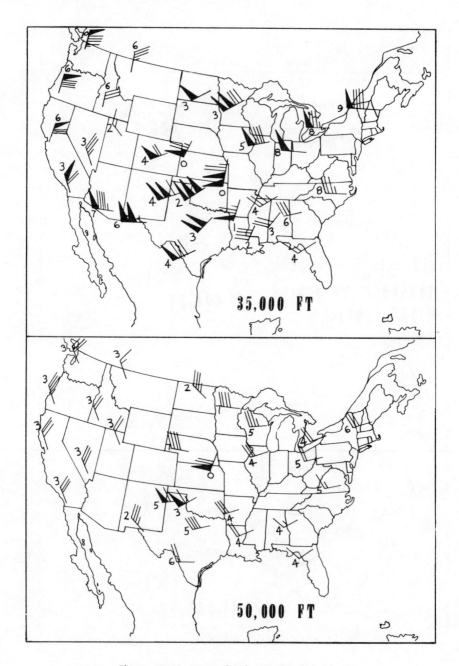

Figure 12–15 Upper-levels Winds Aloft Chart.

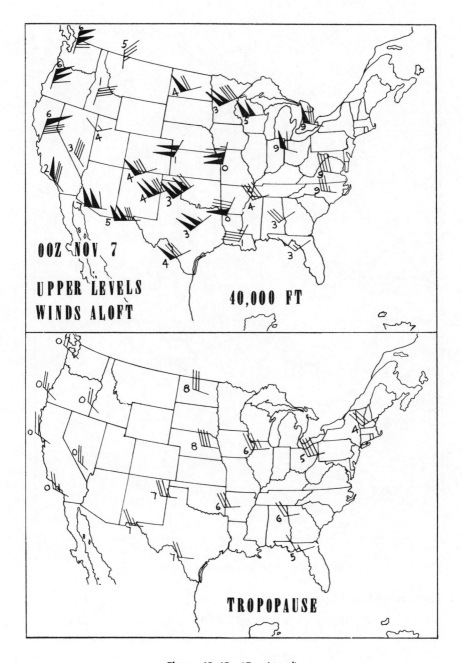

Figure 12–15   (Continued)

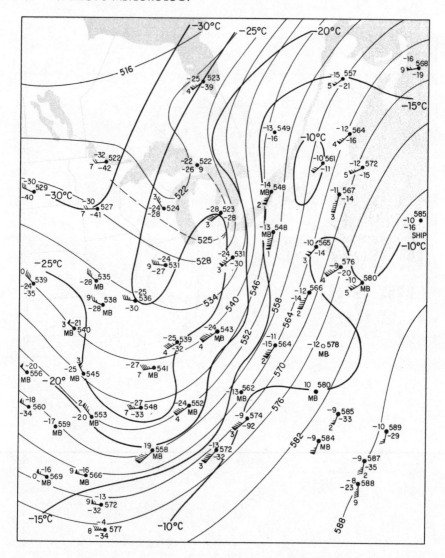

Figure 12–16   Constant-pressure 500-mb Chart.

Constant-Pressure-Chart Levels

The rate in change of pressure with altitude is controlled by the rate in change of temperature with altitude. For example, if the temperature of the air decreases faster than it should, the pressure level of 850 mb, which would be approximately 5000 ft in standard atmosphere, may be at 4600 ft or lower in the cold air. When the temperature lapse rate is less than standard, the 850-mb level may be 5400 ft or higher in the warmer than standard air. The heights of a pressure level are also influenced by the variations in pressure above the level and will influence the sea-level pressure. A constant-pressure chart for the 850-mb level depicts the amount of altitude variation of the 850-mb constant-pressure level in the atmosphere. Wherever the pressure level varies 60 meters (approximately 200 ft), a new contour line is drawn. The same type of chart can be drawn for any pressure level desired. However, the standard levels with which the pilot is concerned are:

850-Millibar Chart   In the standard atmosphere the 850-mb level is 1457 meters (4781 ft). For planning flights at approximately 5000 ft, this chart is the most useful. (One meter = 32808 ft.)

700-Millibar Chart   In the standard atmosphere the 700-mb level is 3012 meters (9882 ft). This chart is used for planning flights near the 10,000-ft level.

500-Millibar Chart   In the standard atmosphere the 500-mb level is 5574 meters (18,289 ft). This chart is used for flight planning near the 18,000-ft level. At this altitude, the atmospheric pressure decreases to approximately one-half the mean sea-level pressure value.

300-Millibar Chart   In the standard atmosphere the 300-mb level is 9164 meters (30,065 ft). The 300-mb chart is significant in that the strong winds of the jet stream, the associated wind shear and clear-air turbulence, and the tops of thunderstorms will frequently appear at this level. It is also used for the interpolation of temperatures and winds for flights between the 18,000-ft and 30,000-ft levels.

(There are also 200-mb [38,000 ft] and 100-mb [54,000 ft] charts, but they are used mainly for research purposes.)

Constant-Pressure Chart Station Model*

The station model used on constant-pressure charts is much simplified from the international station model (Figure 12–3) but uses the same format and symbology:

*See Figure 12–18 for station models used on the National Meteorological Center constant-pressure charts as transmitted on the national facsimile network.

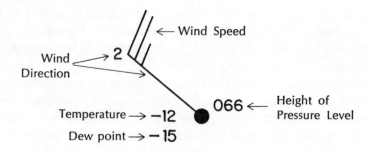

Wind—320° true, 25 knots
Temperature—⁻12°C.
Dew Point—⁻15°C.
Pressure Level—3066 meters mean
(700 mb)   sea level (MSL)
                 (10,059 ft)

- Wind direction is plotted to the nearest 10°.
- Wind speed is plotted to the nearest 5 knots.
- A half-barb is 5 knots, a full barb is 10 knots, a pennant is 50 knots.
- The wind blows from the barbs toward the station circle. If there is a number beside the barbs, it is the middle number of a three-digit symbol of the wind direction. (An east wind would be 9 from 090.)
- If the observed wind is at the exact height for a given pressure altitude, the barbs are solid, thus:

- If the wind is *near* the exact pressure altitude, the barbs are broken, thus:

- The height plotted at the station circle is the elevation at which the balloon or rawinsonde reached the pressure surface that the chart represents. Heights are plotted on the 850-mb and 700-mb charts in meters with the first digit of the whole value omitted. A plotted height of 503 on the 850-mb chart, for example, represents a height of 1503 meters. Heights on the charts at the 500-mb surface and higher are plotted in tens of meters, and a zero must be added to obtain the whole value. On the 200-mb and 100 mb charts, the first digit (always 1) also is omitted. For example, a plotted height of 209 on the 200-mb chart represents a height of 12,090 meters.
- Station circles on the 850-mb, 700-mb, and 500-mb charts are shaded at stations where the spread between the temperature and dew point

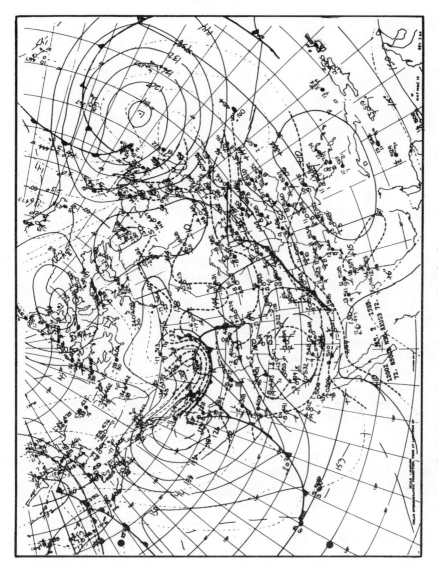

Figure 12-17   Constant-pressure 850-mb Chart.

PIBAL OBSERVATION

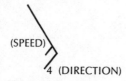

(SPEED)

4 (DIRECTION)

ON THE 700-MB CHART

Height—3000 meters
Wind direction—140°
Wind speed—15 knots

RAWINSONDE OBSERVATION

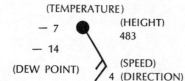

(TEMPERATURE)
— 7   (HEIGHT)
           483
— 14
(DEW POINT)   (SPEED)
           4 (DIRECTION)

AT THE 700-MB LEVEL

Height—3,483 meters
Wind direction—140°
Wind speed—15 knots
Temperature— —7°C.
Dew point— —14°C.

Figure 12–18   Explanation of Pibal and Rawinsonde Observation.

at the height of the pressure surface is 5°C. or less. Above the 500-mb surface, all station circles are shaded because this makes location of stations easier. This point is emphasized because many pilots know that the symbol in the station circle on surface weather charts does indicate the amount of total sky cover, complete shading of the circle signifying an overcast sky. However, the shading of station circles on constant-pressure charts is not related to the amount of sky cover.

■ Temperature and dew point are plotted to the nearest whole degree Celsius.

Reading a Constant-Pressure Chart

Figure 12–17 is a constant-pressure chart for 850 mb. Here is how to read it:

■ *Solid (unbroken) lines* are lines of equal heights, referred to as *contours.* They are drawn for intervals of 60 meters on charts below the 300-mb surface, and for intervals of 120 meters on the 300, 200, and 100-mb charts. The height lines may be used in the same way as isobars on surface weather charts. Above the surface friction effect, the wind blows generally parallel to the height lines, and wind speeds increase as the spacing between the height lines decreases.

■ *Broken lines with long dashes,* also lines of equal heights, sometimes are drawn for intermediatè intervals (every 30 meters below 300 mb) in order to provide better definition of the hills and valleys of pressure.

■ *Short dashed lines,* called *isotherms,* connect points of equal temperature. They normally are drawn for 5°C. intervals. Many local weather offices color isotherms in red for easy identification.

- *Dotted lines* are lines connecting points of equal wind speeds regardless of wind direction. These are called *isotachs*. Isotachs are drawn only on the 300, 200, and 100-mb charts. Intervals are 25 knots for wind speeds up to 150 knots and 50 knots for speeds above 150 knots.

- *A heavy broken line with arrowheads* indicates an *axis of maximum wind*, and is generally referred to as an *axis of the jet stream*. These are indicated only on the 300, 200, and 100-mb charts, except that once a day the axes of maximum wind at the 500-mb surface are transmitted in a chart projection which covers most of the Northern Hemisphere.

- An explanation of the PIBAL and rawinsonde observation is also given on Figure 12–18.

13 AVIATION WEATHER FORECASTS

GENERAL

"What is the weather going to be on this flight?" is the question every pilot asks himself—or should ask —before each flight.

The prudent aviator planning a flight is obviously very much concerned with the forecast weather conditions along his proposed route. So after he has consulted the surface maps, which tell him something of both past and present weather, and then the sequence reports, which are the most up-to-date information on *current* weather, he should then examine the forecasts. There are many forecasts, but the following ones will be covered here in some detail:

- Terminal forecasts
- Area forecasts
- Winds aloft forecasts
- Surface prog charts

- Hurricane advisories
- Severe weather outlooks
- Severe weather forecasts
- Constant-pressure prog charts
- In-flight advisories

A new pilot may be confused by the different words—predictions, forecasts, advisories, outlooks, and prognoses. A pilot will encounter all of them, and while there are slight variations in meanings among them, all are intended to convey that some *future* condition of the weather is expected.

TERMINAL FORECASTS

There are two main types of terminal forecasts—Weather Bureau and military. However, their makeup and content are the same. They are prepared for approximately 500 civilian and military terminals in the United States. Weather Bureau terminal forecasts are issued for 12-hr (over Circuit A) and 24-hr (over Service C) periods. Military terminal forecasts are issued four times a day over the SAUS network.

The Weather Bureau terminal forecast is issued every four hours. The one for 12-hr periods is distinguished by an "FT1" heading while the 24-hr forecast uses an "FT2" heading. These are shown as Figures 13–1 and 13–2 respectively.

The military use two types of terminal forecasts, the internationally sponsored TAFOR (Terminal Aviation FORecast) which is coded, and the U.S.-developed PLATF (Plain LAnguage Terminal Forecast) which is not. The TAFORs are more complete than either the FTs or the PLATFs, but since they are transmitted only by selected U.S. Air Force stations and not available to commercial or private pilots, they will not be covered herein.

Both the Weather Bureau and the military forecasts are transmitted over the teletype network in an abbreviated form which is peculiar to aviation. As can be noted in Figures 13–3 and 13–4, they contain symbols and contractions found on the sequence reports.

Whenever it is advisable for reasons of safety or efficiency, *amended* terminal forecasts are issued. A few such occasions might be:

(1) a change in the forecast ceiling;
(2) a forecast wind which is considerably in error;

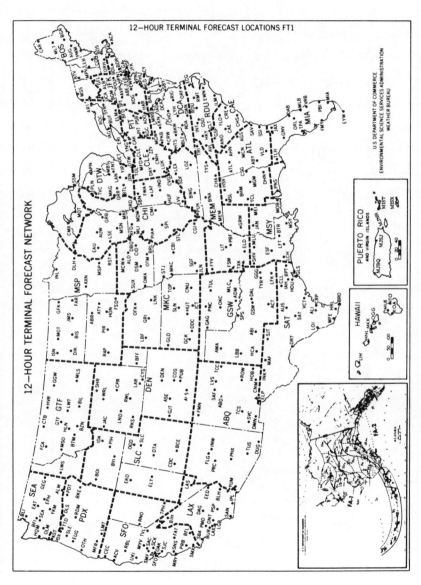

Figure 13-1 Aviation Forecast Centers and Areas, Plus 12-hour Terminal Forecast Network. (Courtesy U.S. Weather Bureau.)

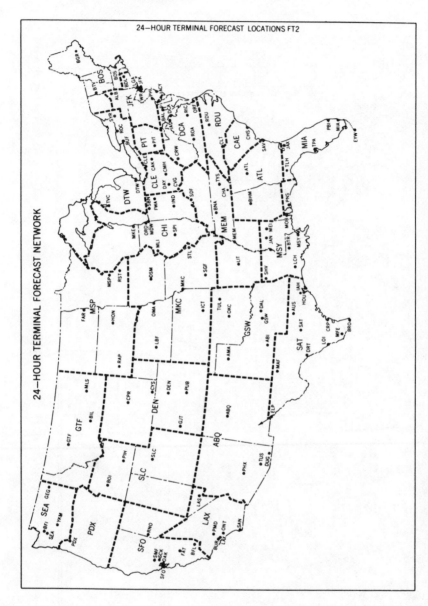

Figure 13–2 Twenty-four Hour Terminal Forecast Network. (Courtesy U.S. Weather Bureau.)

(3) a thunderstorm that was forecast but does not occur—or vice versa;

(4) a significant change in the forecast visibility.

The following are examples of the Weather Bureau's FT forecast (Figure 13–3) and the military FTUS forecast (Figure 13–4A) (also identified as the PLATF).

FT1 DCA 221045
11Z - 23Z MON
FT1   —FORECAST, TERMINAL TYPE, 12 HOUR FORECAST (24 HOUR FORECAST)
DCA   —STATION IDENTIFICATION LETTERS OF FORECAST CENTER FOR REGION
22   —DAY OF MONTH
1045   —FORECAST DISTRIBUTION TIME (GMT)
11Z   —BEGINNING OF FORECAST PERIOD 1100ZULU (GMT)
23Z MON   —END OF FORECAST PERIOD (SUNDAY)
PHL C1ØØ⊕ 3413. 15Z C4ØⰍ 3413 OCNL RW.⁻ Ø3Z C35Ⰽ8ØⰍ 3615 CHC OCNL RW.–.
ALB C4ØⰍ8Ø⊕ OCNL C18⊕5RW.– Ø6Z C35Ⰽ1ØØ⊕ 3610.

The following the the decodes of the terminal forecasts above for Philadelphia and Albany:

PHL   PHILADELPHIA, PENNSYLVANIA
C1ØØ⊕   CEILING OVERCAST CLOUD LAYER AT 10,000 FT ABOVE TERRAIN
3413   WIND FROM 340° AT 13 KNOTS
15Z   FORECAST COMMENCING 1500Z
C4ØⰍ   CEILING BROKEN CLOUD LAYER AT 4000 FT ABOVE TERRAIN
3413   WIND FROM 340° AT 13 KNOTS
OCNL   OCCASIONAL
RW⁻   LIGHT RAIN SHOWERS
Ø3Z   FORECAST COMMENCING 0300Z
C35Ⰽ   CEILING BROKEN CLOUD LAYER AT 3500 FT ABOVE TERRAIN
8ØⰍ   OVERCAST CLOUD LAYER AT 8000 FT ABOVE TERRAIN
3615   WIND FROM 360° AT 15 KNOTS
CHC   CHANCE OF
OCNL   OCCASIONAL
RW⁻   LIGHT RAIN SHOWERS

ALB   ALBANY, NEW YORK
C4ØⰍ   CEILING BROKEN CLOUD LAYER AT 4000 FT ABOVE TERRAIN
8Ø⊕   OVERCAST CLOUD LAYER AT 8000 FT ABOVE TERRAIN
OCNL   OCCASIONAL
C18⊕   CEILING OVERCAST CLOUD LAYER AT 1800 FT ABOVE TERRAIN
5   VISIBILITY 5 MILES
RW⁻   LIGHT RAIN SHOWERS
Ø6Z   FORECAST COMMENCING 0600Z
C35Ⰽ   CEILING BROKEN CLOUD LAYER AT 3500 FT ABOVE TERRAIN
1ØØ⊕   OVERCAST CLOUD LAYER AT 10,000 FT ABOVE TERRAIN
3610   WIND FROM 360° AT 10 KNOTS

Figure 13–3   Weather Bureau Terminal Forecast Code.

```
FTUS 24 KMEM 130520
05Z MON 05Z TUE
TVC 3⓪C8⊕3RF Ø71Ø VRBL C3⊕1RF CHC TSTM AFT 2ØZ..
MEM 1ØØ⓪3ØØ⓪ 151Ø. Ø8Z C4Ø⓪1ØØ⊕ 151Ø CHC RW.⁻ 11Z C1Ø⊕7
181Ø OCNL RW⁻
```

Here are the decodes for these terminal forecasts:

TVC	TRAVERSE CITY, MICHIGAN
3⓪	SCATTERED CLOUD LAYER AT 300 FT ABOVE TERRAIN
C8⊕	CEILING OVERCAST SKY CONDITION AT 800 FT ABOVE TERRAIN
3	VISIBILITY 3 MILES
RF	RAIN AND FOG
Ø71Ø	WIND FROM 070° AE 10 KNOTS
VRBL	SKY CONDITIONS VARIABLE
C3⊕	CEILING OVERCAST SKY CONDITION AT 300 FT ABOVE TERRAIN
1	VISIBILITY 1 MILE
RF	RAIN AND FOG
CHC	CHANCE OF
TSTM	THUNDERSTORM
AFT	AFTER
2ØZ	2000Z
MEM	MEMPHIS, TENNESSEE
1ØØ⓪	SCATTERED CLOUD LAYER AT 10,000 FT ABOVE TERRAIN
3ØØ⓪	BROKEN CLOUD LAYER AT 30,000 FT ABOVE TERRAIN
151Ø	WIND FROM 150° AT 10 KNOTS
Ø8Z	FORECAST COMMENCING AT 0800Z
C4Ø⓪	CEILING BROKEN CLOUD LAYER AT 4000 FT ABOVE TERRAIN
1ØØ⊕	OVERCAST CLOUD LAYER AT 10,000 FT ABOVE TERRAIN
151Ø	WIND FROM 150° AT 10 KNOTS
CHC	CHANCE OF
RW⁻	LIGHT RAIN SHOWERS
11Z	FORECAST COMMENCING AT 1100Z
C1Ø⊕	CEILING OVERCAST CLOUD LAYER AT 1000 FT ABOVE TERRAIN
7	VISIBILITY 7 MILES
181Ø	WIND FROM 180° AT 10 KNOTS
OCNL	OCCASIONAL
RW⁻	LIGHT RAIN SHOWERS

Figure 13–4A U.S. Military Terminal Forecast (FTUS) Code.

AREA FORECASTS

Area forecasts cover a relatively small geographic area and are sent by teletype (Service A) four times daily. They give specific information about flight conditions for the next 12-hour period. While terminal forecasts describe weather at specific airfields, area forecasts are intended as en route forecasts. (For those airfields which have no terminal forecasts [FTs], the area forecast must be used.)

Area forecasts can be recognized on the teletype by the designator FA, and they include forecasts of cloud tops, icing, turbulence, thunderstorms, and other hazards in the following sequence:

- Heading
- Forecast area
- Height above sea level unless noted
- Synopsis
- Clouds and weather
- Icing (and freezing level[s])
- Turbulence (when applicable)
- Outlook

Area forecasts are intended primarily for short- and medium-range flights, and are made for 25 forecast areas in the United States and for Alaska and Hawaii.

Figure 13–5 gives a sample area forecast and its decode.

Heading

The heading identifies the forecast center, the type of forecast to follow, the scheduled filing time, and the valid period.

Forecast Area

This section identifies the geographic area covered by the forecast.

Synopsis

In this section, a synopsis of the general weather to be expected in the terminal area for the forecast period is given, as described in the following sections.

Clouds and Weather   Primary attention in this section is given to a description of the amount and height of sky cover, cloud tops, location and movement of weather-producing fronts as necessary, surface visibility, state of weather and obstructions to vision, surface wind, and other information needed to describe flying conditions expected during the first 12-hour period. Abbreviated plain language normally is used, but authorized weather symbols are employed to describe the amount and height of sky cover, visibility, and obstructions to vision.

Sky Cover   Sky cover is stated solely on the basis of the expected amount of coverage by individual layer, instead of by the summation principle used in sequence reports and terminal forecasts. Adjacent broken or overcast layers with bases 5000 ft or higher above ground and with less than 3000 ft of vertical separation are considered as one layer.

1214 CYQX 34Ø18/32 8ØØØ 51DZ 3STØ5 8SC15 TEMPO 32ØØ 58RA 5STØ5
8SC15
1212 CYYZ 1ØØ15/25 9999 TEMPO 48ØØ 8ØRASH 8NS15 GRADU 2Ø23
9991Ø 4ØØØ 8ØRASH 4STØ6 8NS1Ø TEMPO 16ØØ 95TS 8NSØ5 GRADU
Ø3Ø6 26Ø15 9999 TEMPO 8ØRASH 8SC 2Ø

1214	FORECAST COVERING 1200Z TO 1400Z
CYQX	GANDER, NEWFOUNDLAND
34Ø18/32	WIND FROM 340° AT 18 KNOTS WITH MAXIMUM GUSTS TO 32 KNOTS
8ØØØ	VISIBILITY 80 KM
51DZ	LIGHT DRIZZLE
3STØ5	3/8 STRATUS AT 500 FT ABOVE TERRAIN
8SC15	8/8 STRATOCUMULUS AT 1500 FT ABOVE TERRAIN
TEMPO	TEMPORARILY
32ØØ	VISIBILITY 32 KM
58RA	VERY LIGHT RAIN
5STØ5	5/8 STRATUS AT 500 FT ABOVE TERRAIN
8SC15	8/8 STRATOCUMULUS AT 1500 FT ABOVE TERRAIN
1212	FORECAST COVERING 1200Z TO 1200Z
CYYZ	TORONTO, CANADA
1ØØ15/25	WIND FROM 100° AT 15 KNOTS WITH MAXIMUM GUSTS TO 25 KNOTS
9999	VISIBILITY 10 KM OR MORE
TEMPO	TEMPORARILY
48ØØ	VISIBILITY 48 KM
8ØRASH	RAIN SHOWERS
8NS15	8/8 NIMBOSTRATUS AT 1500 FT ABOVE TERRAIN
GRADU	GRADUALLY BECOMING
2Ø23	BETWEEN 2000Z AND 2300Z
9991Ø	WIND VARIABLE AT 10 KNOTS
4ØØØ	VISIBILTY 40 KM
8ØRASH	RAIN SHOWERS
4STØ6	4/8 STRATUS AT 600 FT ABOVE TERRAIN
8NS1Ø	88 NIMBOSTRATUS AT 1000 FT ABOVE TERRAIN
TEMPO	TEMPORARILY
16ØØ	VISIBILITY 16 KM
95TS	THUNDERSTORM
8NSØ5	8/8 NIMBOSTRATUS AT 500 FT ABOVE TERRAIN
GRADU	GRADUALLY BECOMING
Ø3Ø6	BETWEEN 0300Z AND 0600Z
26Ø15	WIND 260° AT 15 KNOTS
9999	VISIBILITY 10 KM OR MORE
TEMPO	TEMPORARILY
8ØRASH	RAIN SHOWERS
8SC2Ø	8/8 STRATOCUMULUS AT 2000 FT ABOVE TERRAIN

Figure 13–4B   Two Canadian Terminal Forecasts with Their Translations.

Cloud Heights   Ordinarily heights are stated in hundreds of feet, using intervals of 100 ft for heights up to 3000 ft above ground in plains and valley areas. Above 3000 ft, intervals of 500 to 1000 ft are used. Heights may be with reference to mean sea level or above ground, whichever most effectively conveys the information. However, heights are, in each instance, specifically identified as to the height reference. Ceiling heights are always with reference to above ground as the "C" implies, and no other identification is necessary.

Cloud Tops   The height of cloud tops is stated for cloud layers with bases 20,000 ft mean sea level or lower. Cloud-top information is written in hundreds of feet, using mean sea level as a reference plane, and is identified by the word "TOPS."

Visibility   Surface visibility of more than eight statute miles is omitted. When visibilities are forecast to be six statute miles or less, the state of weather and obstructions to vision will be included. This rule is the same as for terminal forecasts. Flight or slant visibility is not included in the routine forecasts, but is sometimes implied by the forecast clouds, weather, and surface prevailing visibility.

Weather and Obstructions to Vision   Occurrences of weather and obstructions to vision are written in contracted form (for example, SNW, DRZL) when included with a plain language statement. They are indicated in symbol form (S, L, etc.), using the plus or minus signs in accordance with aviation reporting procedures, when written as part of a complete symbolic group.

Surface Winds   Surface winds are stated in symbolic form for any area where winds are expected to reach sustained speeds of 25 knots or more. Directions are stated with reference to true north and speeds are given in knots. The contraction SFC WND will precede the direction and speed group. Gusty surface winds included in the plain language or abbreviated text portion of the forecast are denoted by the term GUSTY or GUSTS TO . . . or, when part of the complete symbolic group, by the form 25G, 25G40, or G65.

Icing   This section, identified by the contraction ICG, includes a statement of expected icing conditions plus the height of the freezing level. Types of icing (clear, rime, or mixed) are indicated by the contractions CLR, RIME, or MXD. Icing intensities are described as light, moderate, or heavy by the contractions LGT, MDT, and HVY. Sometimes contractions such as ICGIC, ICGIP, and ICGICIP, meaning respectively "icing in clouds," "icing in precipitation," and "icing in clouds and in precipitation," are used in combination with icing type and intensity.

Qualifying terms such as "probably," "likely," and "locally" are used when these add to the value of the forecast, such as: MXD ICGIC LKLY. A forecast of no icing is written NONE. Even when no icing is forecast, the expected freezing level is indicated. Heights of icing and the freezing level are stated in hundreds of feet above mean sea level.

Turbulence

This section, identified by the contraction TURBC, is included when turbulence of any intensity sufficient to affect safety of aircraft is expected. The terms LGT, (light), MDT, (moderate), SVR, (severe), and expected. The terms LGT (light), MDT (moderate), SVR (severe), and EXTRM (extreme) are used to describe intensity. If no important turbulence is expected, this section is omitted. Heights are given in hundreds of feet above mean sea level.

Outlook

This section contains a brief statement of conditions expected in the 12-hour period immediately following the forecast period. Emphasis is on weather conditions significant to flight planning.

Here is a typical area forecast:

FA CLE 181845

19Z FRI-07Z SAT

OHIO W PA W NY

HGTS ASL UNLESS NOTED

SYNOPSIS. CDFNT AT 19Z FRI WINDSOR-CINCINNATI LN MOVG TO NR NIAGARA FALLS-PARKERSBURG LN AT 00Z AND TO NR UTICA-HARRISBURG LN BY 07Z

CLDS AND WX. ALG AND TO 150 MI E OF CDFNT MSTLY C25–40 2115G45 OCNL C12 2RW-ISOLD TSTMS. TOPS GENLY 80–120 BLDUPS TO 350. FTHR E OF FNT THRU W NY AND W PA GENLY 40–60 250 WITH INCRG SWLY SFC WNDS. W OF CDFNT MSTLY C15–30 60–70 BCMG VRBL TO OVER W OHIO BY SNST AND E OHIO BY 07Z.

ICG. LCL MDT RIME ICGIC. FRZLVL 80–90 LWRG TO 50–60 W OHIO 07Z

TURBC. FQTLY MDT BLO 100. LCL SVR SFC–50 DUE STG LOW

Figure 13–5   Area Forecast Code.

LVL WNDS. SVR ALL LVLS IN AND VCNTY HVYR RAIN SHWRS OR
TSTMS. PSBL MDT OR GRTR CAT 240–280 OHIO MOVG EWD
INTO W PA W NY 07Z

OTLK 07Z– 19Z SAT. OVR W PA W NY MSTLY C15–30 V BCMG
GENLY C30–50 OCNL ENTR AREA BY SNRS. CHC OF A FEW RAIN
SHWRS OR MXD RAIN AND SNW SHWRS OVR EXTRM NW PA
W NY BY 19Z

Here is the decode:

Forecast, area type, issued by the Cleveland Forecast Office and
distributed on the teletype circuits on the 18th day of month at
1845Zulu(GMT). The forecast period commences at 1900Zulu on
Friday and ends at 0700 Zulu on
Saturday

The area forecast will cover Ohio, Western Pennsylvania, and Western
New York.

Heights mentioned in forecast are above sea level unless noted.

Synopsis. Cold front at 1900Zulu Friday Windsor-Cincinnati line
moving to near Niagara Falls-Parkersburg line at 0000Zulu and to near
Utica-Harrisburg line by 0700Zulu.

Clouds and weather. Along and to 150 miles east of cold front mostly
ceiling 2500-ft to 4000-ft broken, wind from 210° at 15 knots with
gusts to 45 knots. Occasional ceiling 1200 broken and 2 miles of
visibility in light rain showers of isolated thunderstorms. Tops generally
8000 ft to 12,000 ft with build-ups to 35,000 ft. Further east of front
through Western New York and Western Pennsylvania generally 4000
ft to 6000 ft scattered and 25,000 ft thin broken with increasing
southwesterly surface winds. West of cold front mostly ceiling 1500
ft to 3000 ft overcast topped at 6000 ft to 7000 ft becoming variable
to broken over Western Ohio by sunset and Eastern Ohio by
0700Zulu.

Icing. Local moderate rime icing in clouds. Freezing level 8000 ft to
9000 ft lowering to 5000 to 6000 ft over Western Ohio by 0700Zulu.

Turbulence. Frequently moderate below 10,000 ft. Local severe from
the surface to 5,000 ft due to strong low level winds. Severe all
levels in and vicinity heavier rain showers or thunderstorms. Possible
moderate or greater clear air turbulence 24,000 ft to 28,000 ft over
Ohio moving eastward into Western Pennsylvania and Western New
York by 0700Zulu.

Outlook from 0700Zulu to 1900Zulu Saturday. Over Western
Pennsylvania and Western New York mostly ceiling 1500 ft to 5000 ft
broken occasional scattered entire area by sunrise. Chance of a few
rain showers or mixed rain and snow showers over extreme
Northwestern Pennsylvania and Western New York by 1900Zulu.

Figure 13–5 (Continued)

WINDS ALOFT FORECASTS

Winds aloft forecasts are made every six hours for 89 selected locations in the United States and are transmitted on Service A. They carry the designator FD. These forecasts are computer-produced by the National Meteorological Center twice a day (0550Z and 1750Z), and they are manually produced by the Flight Advisory Weather Service twice a day (1150Z and 2350Z). All four winds aloft forecasts issued daily are for a 12-hour period, but those which are computer-produced are 6-hour forecasts. Figure 13–6 is a sample winds aloft forecast and its code.

Each winds aloft forecast includes the expected wind direction and speed for 3000 ft (for stations having a terrain elevation of 2000 ft or lower), 5000 ft (for stations having a terrain elevation of 4000 ft

```
FD WBC 292350
00–06Z

LVL 3000 5000FT 7000 10000FT 15000FT 20000FT 25000FT
GRI 2418 2422+17 2427 2431+10 2540+02 2549–15 2558–29
DEN            2521 2529+08 2444–04 2456–16 2468–30
06–12Z

LVL 3000 5000FT 7000 10000FT 15000FT 20000FT 25000FT
GRI 2123 2224+19 2327 2429+12 2535+00 2544–13 2554–28
12–18Z
```

FD	—Letters used to identify winds aloft forecast.
WBC	—Weather Bureau Central (originator of forecasts).
29	—29th day of month.
2350	—Beginning transmission time (GMT).
00–06Z	—Forecast period.
LVL	—Altitudes above sea level.
GRI/DEN	—Station identification letters. (Green Island, Nebr.; Denver)
2418	—240°/18 knots/3,000′.
2422+17	—240°/22 knots/+17°C/5,000′.
2427	—240°/27 knots/7,000′.
2431+10	—240°/31 knots/+10°C/10,000′.
2540+02	—250°/40 knots/+2°C/15,000′.
2549–15	—250°/49 knots/−15°C/20,000′.
2558–29	—250°/58 knots/−29°C/25,000′.

Figure 13–6 Winds Aloft Forecast Code.

or lower), 10,000 ft, 15,000 ft, 20,000 ft, and 25,000 ft. All heights are in thousands of feet above mean sea level. Temperature forecasts are added to all wind forecasts above 3000 ft, except no temperature forecasts are appended to the 5000-ft wind forecast when this is the lowest-level forecast.

Direction

The wind direction is in reference to true north and is reported in two digits. For example, 180° is reported as 18.

Speed

The windspeed is reported in knots. Speeds less than 10 knots are shown as Ø8, Ø6, etc. If the speed is expected to be less than 5 knots, the group 99ⓞⓞ is inserted instead of the direction and speed. This group is known as "light and variable."

SURFACE PROG CHART

The surface prog chart issued by the NMC provides a general picture of surface fronts, high and low pressure centers, wind patterns, low and middle clouds, precipitation areas which can be expected in the next 24-hour period of time. (The Weather Bureau also issues 36- and 48-hour progs.) Figure 13–7 is an example of a 24-hour surface prog chart.

These forecasts are used as forecasting aids at local weather stations. (The local forecast, however, remains the responsibility of the individual issuing it.)

The surface progs are transmitted over facsimile circuits to local weather stations. Many of the forecast charts are *composite,* containing the forecast position of 500-mb contour lines in addition to surface isobars. Composite charts indicate isobars as solid black lines and contour lines as dashed black lines.

Approximately seven forecast surface and composite charts are transmitted by facsimile process each day. Isobars are drawn at 4-mb intervals and labeled in multiples of four. Standard frontal symbols are used, and the fronts are coded as to type and intensity. The centers of high (H) and low (L) pressure are identified, with central pressures labeled in tens and units of millibars. The forecast direction of movement of the pressure centers is normally indicated by an arrow, with

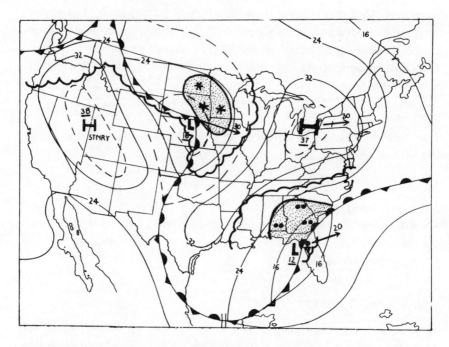

Figure 13–7   Twenty-four Hour Surface Prog Chart.

the speed of movement shown in knots. The cross-hatched areas on the chart indicate regions where an overcast cloud cover or an overcast with a few breaks will occur at the verifying time (VT) of the prognosis. Stippled areas indicate where precipitation is expected. Areas of overcast and precipitation are usually colored at the local weather station. Scalloped lines outline predicted areas of low and middle clouds.

HURRICANE ADVISORIES

While most pilots will rarely, if ever, fly in a hurricane area, they will occasionally see hurricane advisories on both Service A and C. Such advisories carry a WH designator. Whenever a hurricane is threatening an area, advisories are transmitted four times daily in plain language. They include the location, extent, intensity, and movement of the hurricane, as well as a forecast of these general features. Specific information on aviation weather hazards associated with the hurricane is contained in appropriate Area and Terminal Forecasts and in In-Flight Weather Advisories.

The greatest use of Hurricane Advisories is for the evacuation of aircraft from threatened locations.

An example of an abbreviated Hurricane Advisory follows:

> WH MIA 181010
> HURCN IONA AT 05E CENTRD 29.4N
> 75.2W OR 300 MI E OF JACKSONVILLE FLA
> EXPCD TO MOV NW ABT 12 KT. MAX WNDS
> 110 KT OVR SML AREA NEAR CNTR AND
> HURCN WNDS WITHIN 55–75 MI.

SEVERE WEATHER OUTLOOKS

During the thunderstorm season, an outlook for thunderstorm activity in the contiguous United States is transmitted on Service A early each morning by the Severe Local Storm (SELS) Forecast Center. It carries the designator AC. This outlook, stated in rather general terms and in plain language, describes the prospects of severe weather during the next 24 hours. The term SVR LIMITS, meaning conditions requiring the issuance of detailed severe weather forecasts, is used frequently in the text. Aviation interests use this information on the probability of severe local storms in operational planning.

A sample Severe Weather Outlook (AC) is shown below:

100
AC MKC 051040
MKC AC 050440C
WIDESPREAD TSHWR ACTVY EXPCD TDA ALG FRONTAL BNDRY FM
WRN APLCNS THRU OHIO VLY ALG NRN ILL AND WWD INTO
CNTRL IA NRN NEB WRN DKTS AND ERN MONT ALG ERN RCKYS.
THE USUAL AMS TSHWRS WL AGN OCR ALG THE CSTL CAROLINAS
AND GLF CST. CNVGG PATN OF JTSTR FM GRTLKS EWD INDCS
TSHWR PCPN IN ILL IND AND OHIO WL BE PRINLY OVRNG AND
RATHER COPIOUS BUT BLO LIMITS. MOST LKLY AREA TO
ATTAIN SVR LMTS WL BE ALG INSTBY LINE FRMG EARLY AFTN ALG
E SLP RCKYS FM CNTRL MONT SWD INTO CNTRL WYO WITH
LINE MOVG INTO CNTRL DKTS AND CNTRL NEB. MAGOR.

SEVERE WEATHER FORECASTS

Severe Weather Forecasts are also issued by the Severe Local Storm Center (SELS) in Kansas City, Missouri. They carry a WW heading identification. These plain-language forecasts are of particular interest to air traffic service personnel, local weather offices, and pilots. The content of SELS forecasts is reflected in Terminal and Area Forecasts and in In-Flight Weather Advisories, and are sent on the Weather Bureau teletype service.

The general format of Severe Weather Forecasts is shown in the following example:

100
WW MKC 050140
MKC FCST NR 423 050140
AREA ONE . . . . SVR TSTM FCST
A . . . . ALG AND 60 MI EITHER SIDE OF A LINE FM 30 MI
SW OF MASON CITY IA TO 20 MI S OF PEORIA ILL. VALID
CURRENT TIL 0600Z. PUB FCST ISSUED.
B . . . . SCTD SVR TSTMS WITH HAIL TO ¾ INCH DIAM EXTRM
TURBC AND MAX SFC WND GSTS TO 60 KT. SCTD CBS MAX TOPS
TO 55 THSD FT.
C . . . . SQLN DVLPG IN NWRN IA EXPCD TO MOV SE ABT 35 KT.
REMARKS. THIS FCST REVISED WW 421 AREA TWO
SMITH.

An explanation for each part of the format follows:

AREA (number)—*TYPE OF FORECAST* (Tornado, etc.)—Forecasts for more than one area may be issued under the same heading, in which case the areas are numbered serially within the issuance. The type of forecast follows the numbered area, as TORNADO FCST or SVR TSTM FCST.

A—This paragraph describes the area in which the severe weather conditions are forecast and states the valid time. If a forecast applicable to the general public has been issued, or will be issued, the statement PUB FCST ISSUED or PUB FCST WIBIS (will be issued) is included. If one of these statements is not included, the forecast is of primary concern to aviation only, and the user should guard against any dissemination and unnecessary alarm to the general public.

B—This paragraph states the number of storms expected as a few, scattered, numerous, etc. It also describes the severe conditions expected with respect to hail, turbulence, and surface winds.

C—This paragraph describes the weather system that is expected to cause the severe conditions, and may give details of reported weather of significance.

REMARKS—as appropriate to complete understanding and usage of the forecast.

CONSTANT-PRESSURE PROG CHARTS

Facsimile forecast charts of upper-air conditions are called *constant-pressure prog charts*. Approximately 20 of these charts are transmitted by the NMC each day for altitudes up to 300 mb and for periods of time up to 72 hours. Charts drawn for altitudes below 300 mb usually consist of contour lines, frontal symbols, and labeled centers of high (H) and low (L) elevation. Constant-pressure prog charts at and above 300-mb altitudes normally include additional data for jet stream axes and wind speed (isotachs).

The 36-hour 700-mb prog chart is used for flight planning at altitudes near 10,000 ft. The contour lines on this chart are drawn at 60-meter intervals and labeled in meters with the thousand digit omitted.

Since the wind flows parallel to the contour lines at a speed directly proportional to the distance between contours, values for wind direction and speed may be computed from constant-pressure prog charts. Weather conditions on the charts depict the conditions expected at the specified verifying time (VT), rather than the average weather conditions expected between transmission time and verifying time.

IN-FLIGHT ADVISORIES

While not strictly forecasts, the in-flight advisories which a pilot receives in the air do warn him in advance of unusual or unexpected weather he may encounter. These advisories were discussed more fully in Chapter 11.

FLYING THE WEATHER

INTRODUCTION

In this final part, we will begin to apply the technical meteorological knowledge and physics presented in earlier chapters toward the planning and conduct of actual flight operations. It is one matter to know what clouds are or how fog forms and what thermodynamic factors produce a thunderstorm or how to interpret a map and read a sequence report. It is quite another matter to apply this knowledge to actual flying conditions—to be able to forecast and avoid fog, to be able to circumvent thunderstorms, or to know how to minimize ice formation during flight.

Most aviation accidents occur to pilots with 500 hours or less of flying time—and, of those accidents, many are caused by ignorance of weather phenomena, overconfidence in the face of uncertain weather conditions, or failure to use the weather information that is available, either in planning or en route.

Aviation accident statistics clearly show the wisdom of the old saying, "There are old pilots, and there are bold pilots; but there are no old, bold pilots." Know your meteorology; respect the weather; use the weather equipment and facilities which are available—and you won't become an aviation statistic because of marginal weather conditions.

14 FOG AND FLYING

Fog is one of the most common, and potentially one of the most dangerous weather hazards a pilot will encounter. It is most likely to be encountered in two places: (1) near bodies of water—ocean, lakes, rivers; (2) over and near industrial cities where there are heavy concentrations of smoke. Fog is also more likely during the cold months of the year than during the warm months.

So let us briefly review what we learned about the physics and thermodynamics of fog, types of fog, and fog formation from Part II.

First, there are five kinds of fog: (1) radiation fog, (2) advection fog, (3) up-slope fog, (4) steam fog, and (5) frontal fog. However, only the first two are of primary concern to pilots.

All of the five types consist of minute water droplets suspended in the atmosphere near the earth's surface. Fog is really a stratus cloud that forms at the ground or so close to it as to seriously impair surface visibility.

Since fog is a stratus cloud at or near the surface, it is the product of condensation. For condensation to occur, the air must become saturated. The approach to saturation is accompanied by an increase in relative humidity which is indicated by a reduction of the temperature–dew-point–temperature spread or the temperature–dew-point depression. Saturation (100% relative humidity) refers to pure water. Foreign particles in the free atmosphere—smoke, dust, salt—serve as condensation nuclei. These nuclei may begin to collect moisture at humidities as low as 70%. Therefore, fogs can form and exist, especially in industrial areas, with humidities as low as 80%.

Such nonsaturated air can become saturated and form fog as a result of cooling of the air. Cooling is the most frequent cause of fog formation. It may be due to loss of heat by the ground because of outgoing radiation (radiation or ground fog), loss of heat of warm air flowing over a cooler surface (advection fog), adiabatic cooling of air streaming over rising terrain (up-slope fog).

Fog may form as a result of nonsaturated air becoming saturated through evaporation from water warmer than the air. This fog is known as steam fog. If the evaporation is from rain falling through colder air, the fog is referred to as frontal fog.

Finally, fog rarely forms as a result of a single process. It is usually produced by a combination of processes, although one may be predominant.

GROUND OR RADIATION FOG

Ground or radiation fog is formed by a temperature inversion at the ground as the result of radiational cooling of the land at night. It is generally heaviest at daybreak. This fog will usually disappear during the daylight hours. In the warm months of summer, it will dissipate quickly. In cold or winter months, it may persist longer. However, any cloud layer above a ground fog will retard its dissipation.

While most ground fogs will burn off between 10 A.M. and 12 noon, don't bet on it. If you arrive over a field blanketed by a ground fog, don't circle and wait for it to burn off. Go immediately to your alternate.

ADVECTION FOG

Advection fog forms as the result of moist air blowing over a cool surface, which cools the air to its dew point. For radiation or ground fog to form, little or no wind (2 to 7 knots) is necessary, but wind *is* necessary for advection fog (5 to 15 knots).

The most common advection fog is *coastal* or *sea-breeze* fog. In winter months, land is colder than the adjacent sea. A sea breeze which brings moist sea air over the cold land can quickly bring heavy fog.

Another type of advection fog of concern to pilots is *sea fog,* such as one finds off the California coast in the summer. This occurs because warm ocean air flows across the cold California current.

Such fog can blow in very quickly over a coastal airfield—sometimes in a matter of minutes. So if your destination is in such an area, plan your flight to have a reserve of fuel, and don't hesitate to land at a field on the coastal side of the mountains, or at your preplanned alternate.

FLIGHT PLANNING FOR FOG

Fog normally is not an en route problem but it can be a terminal problem. Since it can form quickly, and at times unexpectedly, a pilot should keep it in mind both during preflight planning and during continuing planning while en route.

For night flights in areas where the air has a high moisture content and light winds, radiation fog is a consideration. In the preflight planning, both the current weather and the forecasts should be checked. Special consideration should be given the reports of the latest observations; and, if the temperature–dew-point spread is narrow—4°F. or less—with calm to light winds, or with stronger winds which have been *decreasing* in speed, careful analysis should be made of the inflight weather reports received while airborne to observe the temperature–dew-point and wind trends. If the temperature is dropping and the wind decreasing to light to calm or remaining light to calm, then radiation fog can be expected before sunrise. If flight is to be conducted within or to a coastal area, advection fog is a possibility. For flights in coastal areas, a pilot should particularly watch the spread. If the air flow is from the warmer to the cooler surface, advection fog can quickly occur. The pilot should also check pressure systems and associated air flow during his preflight planning. He should bear in mind that if the air-mass horizontal pressure gradient is shallow, diurnal surface heating and cooling may cause the surface pressure gradient to be stronger than that of the air mass, with the resulting surface land and sea breezes contrary to the gradient wind. While the pilot should monitor weather broadcasts at all times while en route, it is especially important that he do so on coastal flights. A wind shift could mean trouble.

If a coastal station has been reporting *partial* obscuration in fog during the day and the wind prevails after sunset, it will probably start reporting a complete obscuration.

Steam fog normally is not much of a problem to aviation operations. Up-slope fog can hamper flight operations in the Great Plains area when east winds prevail. Frontal fog is a possibility with almost any front which has associated precipitation.

If preflight planning indicates any possibility of fog forming in the terminal area, the pilot should select an alternate, or alternates. If fog does form in the terminal area, lowering ceiling and/or visibility below approach minimums, the pilot should proceed immediately to the previously selected alternate. As stated earlier, orbiting with the hope that the fog will dissipate or move out is not a prudent action.

Another course of action is to plan your flight with a takeoff time which will get you to your destination after the fog has lifted or dissipated.

Therefore, an aviator should consider the possibility of fog formation at his destination and at alternates during flight planning, especially when the field is on or near the coast or large bodies of water. Obviously, the alternate should be one free of fog or its likelihood. If the destination is near the water with an onshore wind, an inland alternate can be selected which is behind a mountain or ridge. A ridge or range of mountains will act as a barrier and prevent fog from moving inland.

Moreover, a check of the facilities in the weather station can help the aviator anticipate areas and times of fog formation. The teletype sequence reports show the tendency of the temperature–dew-point spread; this tendency may be projected to the time when the spread will become critical. Terminal forecasts indicate the expected ceiling and visibility at the forecast time of fog formation and/or dissipation. These should be checked. Surface weather maps and sequence reports, used together, indicate frontal precipitation areas where fog is likely to form. These two facilities also indicate the direction and velocity of the wind in relation to topography. This relationship is beneficial in predicting areas of advection or up-slope fog formation.

Finally, always consult the forecaster about all probable fog areas since slight changes in temperature, moisture, and wind direction or speed can cause fog to form or to dissipate.

15 THUNDERSTORMS AND FLYING

THUNDERSTORMS

General

In Chapters 7, 8, and 9 the physics and thermodynamics of clouds, air masses, and fronts were explained in detail. In this chapter we will translate that technical information in order to introduce pilots to the flight procedures and techniques regarding thunderstorms.

It has been aptly said that a thunderstorm is a cumulus cloud gone wild. Thunderstorms have also been described as "weather factories"—violent and spectacular storms containing rain, hail, turbulence, lightning, snow, and thunder. The thunderstorm has been the nemesis of many ill-informed aviators who regarded them as playthings. No more serious flying hazard faces a pilot than the thunderstorm. When the pilot understands that a typical three-mile-wide thunderstorm contains ten times the power of a

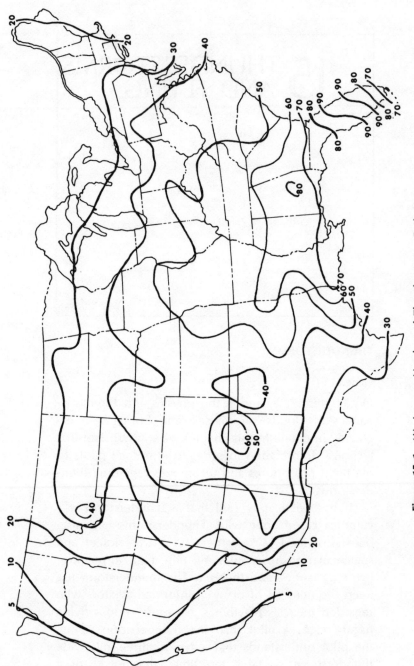

Figure 15–1 (A) Average Number of Thunderstorms Each Year.

Figure 15–1   (B) Cold-front Thunderstorm.

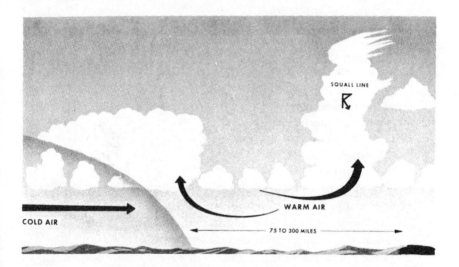

Figure 15–1   (C) Prefrontal Squall Line of Thunderstorms.

Hiroshima atomic bomb, holds a half-million tons of water, can rise to heights of 60,000 ft, can contain hailstones the size of baseballs traveling at 100 knots, can spawn tornadoes, or rain so torrential that visibility outside the cockpit is impossible, he will learn to circum-navigate—and if possible avoid—these awesome devils of the at-mosphere.

It is not the author's intent to make pilots afraid of thunderstorms

—just enormously respectful. Fear will only dull the pilot's responses and reactions when he encounters them. Since something like 44,000 thunderstorms occur daily around the earth, every pilot at some time or other will encounter them. (See Figure 15–1A.) Thus, every pilot should know as much as possible about thunderstorms so that when he does encounter them, he can do so with confidence and respect, not fear.

Types

All thunderstorms have the same physical features. They differ, however, in intensity, degree of development, and associated weather phenomena. Thunderstorms are generally classified according to the way the initial lifting action is accomplished. They are divided into two general groups, frontal thunderstorms and air-mass thunderstorms. These, in turn, are subdivided as follows:

(1) Frontal thunderstorms
    (a) warm-front
    (b) cold-front (see Figure 15–1B)
    (c) prefrontal (squall line) (see Figure 15–1C)
    (d) occluded-front
(2) Air-mass thunderstorms
    (a) convective
    (b) orographic
    (c) nocturnal

For thunderstorms to develop, four factors must be met.
(1) The air must have a high moisture content.
(2) This moist air must be unstable. If the air were stable, stratiform rather than cumuliform clouds would develop.
(3) There must be lifting action. Lifting may be cool, moist air heated by contact with a warm surface. (Thus, thunderstorms form over coastal land surfaces during the day and over adjacent waters at night.) Lifting may also be furnished by orographic lifting. In mountainous areas, thunderstorm activity is frequently found on the windward sides of the mountains.
(4) The cloud formed must have a build-up which goes above the freezing level.

Each thunderstorm progresses through a cycle which consists of three stages: (1) the cumulus stage, (2) the mature stage, and (3) the dissipating or anvil stage. (See Figure 15–2.)

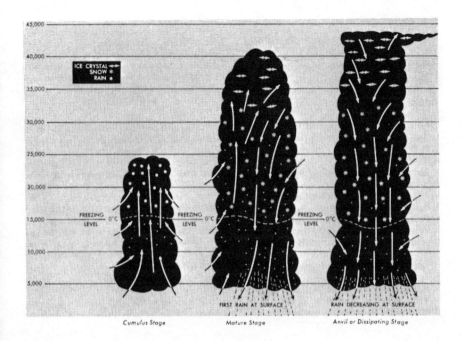

Figure 15–2 Three Stages of a Thunderstorm.

The mature stage is the most dangerous. The storm reaches this stage when rain begins to fall to the ground, and at this time the usual thunderstorm is perhaps 25,000 ft high.

Effects of Thunderstorms

Static warm-front thunderstorms are often difficult to identify because they are obscured by surrounding stratiform clouds. A pilot flying in the relatively smooth air on the cold side of a front into warm air can encounter severe turbulence in a matter of seconds. The best warning is crash static in the earphones in the low and medium frequency bands.

In mountainous areas, where orographic thunderstorms most frequently occur, it is not prudent to fly through the lower portions of the storms since they almost always enshroud the tops of mountains and hills. (See Figure 15–3.)

Hail Hail can be found in three different areas of a thunderstorm: (1) in the rain area within the cloud, (2) falling from the anvil or other overhanging cloud, or (3) as much as four miles from the cloud itself.

Figure 15–3 Orographic Thunderstorm.

Approximately three-fourths of the thunderstorms contain hail which never reaches the ground. No completely accurate method has been found for determining the presence of hail within a thunderstorm, nor has a completely satisfactory method of forecasting the occurrence of hail within a thunderstorm been developed. Hailstones can be quite large—baseball size is rare, but hail of one-half inch size is common. Striking such hailstones can severely damage the wings, tail, radome, etc., or if ingested into a jet engine, can do internal damage. You should be alert, therefore, to the possibility of encountering hail in any thunderstorm, and try and avoid it.

Lightning   According to substantiated reports, lightning which strikes an aircraft in flight occurs in a very small percentage of thunderstorm penetrations. Strikes may occur at any level if conditions are favorable. Lightning is not dangerous to occupants of a metal aircraft, except that it may damage the aircraft itself or the radio equipment. The current in a stroke of lightning may carry 60,000 to 100,000 amperes. Magnetic compasses of aircraft which have been struck by lightning may be 20° to 30° in error. Lightning can also cause a pilot flying instrument to be temporarily blinded. Needless to say, a pilot should avoid the thunderstorm area where lightning is most frequent. In case you are in a thunderstorm where there is considerable lightning, turn up the cockpit lights full bright as early as possible. Some pilots even fly instruments with one eye closed so they won't be flash-blinded in both eyes.

Icing Since ice requires temperatures at or below freezing, the icing level will lower with the freezing level. Inasmuch as the freezing level is also in the zone of greatest frequency of heavy turbulence and heavy rainfall, altitudes within a few thousand feet above or below the freezing level should be regarded as particularly hazardous. Winter, spring, and fall thunderstorms, though not so frequent, more often involve low-level icing. (See Figure 15–4.)

Snow In some thunderstorms, snow will be encountered mixed with supercooled raindrops which can make accumulations of wet snow and rime ice occur quickly on leading edges of wings and tail surfaces and around engine intakes, etc.

Altimeter Errors Rapid and marked surface pressure variations generally occur during the passage of a thunderstorm. The typical sequence of these variations is for an abrupt fall as the storm approaches, an abrupt rise accompanying the rain, and a gradual return to normal as the rain ceases. To a pilot, the importance of such pressure fluctuations is that they may result in significant altitude errors if the altimeter setting used is not current almost to the minute. Remember that any altimeter

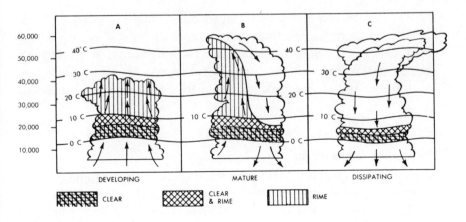

Figure 15–4 Thunderstorm Icing Zones.

setting that you get during or immediately before a thunderstorm is highly unreliable—a 60-ft to 100-ft altimeter error would not be uncommon. See Chapter 4, page 40 regarding altimeter errors.

Tornadoes   With winds as high as 300 knots, the tornadoes spawned by thunderstorms are not meant for airplanes. No pilot has ever flown into a tornado and lived to tell about it. Chapter 7, page 102, gives a detailed description of tornadoes.

Turbulence   Perhaps the worst danger of a mature thunderstorm is its turbulence. There are four types, or conditions, of turbulence:

(1) *Light turbulence*   Light turbulence *may* require the use of seat belts. It produces a variation in airspeed from 5 to 15 knots. Loose objects in the aircraft remain at rest during flight.

(2) *Moderate turbulence*   Moderate turbulence *requires* the use of seat belts. Airspeed is affected, varying from 15 to 25 knots. Objects loose in the aircraft tend to slide or roll.

(3) *Severe turbulence*   Severe turbulence causes the aviator to lose control of the aircraft momentarily. The occupants are thrown violently against seat belts and the seat. Loose objects are tossed about the aircraft, and the airspeed is affected in excess of 25 knots.

(4) *Extreme turbulence*   Extreme turbulence is relatively rare. The aircraft will be tossed about and control is practically impossible. Rapid fluctuation of the airspeed occurs in excess of 25 knots. Such turbulence may cause structural damage to aircraft.

Gusts   The bumps and jolts received during flight through or in the vicinity of thunderstorms can have a considerable effect on an aircraft. Gusts as strong as 27 knots have been measured in research aircraft. Light aircraft can be damaged structurally by such bumps, and even heavy aircraft can be damaged if they are flying at high speeds.

FLIGHT PROCEDURES AND TECHNIQUES

Preparation Before Flight

The meteorological facilities for forecasting and warning about thunderstorms are quite excellent. Radar Summary Charts and radar reports (RAREPs) are helpful in identifying thunderstorm areas, and telling how many there are and how severe they are. A pilot should also check PIREPs, SigMets, and AirMets before he files a flight plan. He should consult with the meteorologist and learn the height of the freezing level as an aid in selecting the proper flight altitude to avoid maximum lightning and icing areas. If a radar scope is available at the weather station, the pilot should check the radar picture to orient to the location and intensity of the storms in the immediate area. The pilot must then decide if it is possible to circumnavigate the storms or to avoid them by over-the-top or under-the-base flight.

Preparation in Flight

When you see a thunderstorm in flight or when you have been warned by the crash static in your earphones that you are approaching a thunderstorm, you have an immediate decision to make. You must decide whether it is possible to circumnavigate the storm or avoid it by going under or, if all routes are closed to you, whether your flight is important enough to attempt penetration.

Circumnavigation

Circumnavigation presents no special flight problems. When you have decided that circumnavigation is possible (as is the case in isolated air-mass storms), you merely alter your course to go around. Since most thunderstorms are less than 50 miles across, slight detour to one side or the other does not appreciably add to either the time or distance of the flight. In case of a line of thunderstorms, it is possible to circumnavigate by flying through "thin spots" between storm centers. In this case, radar is a big help in picking the right hole.

Over-the-top Flight

This is out of the question because of altitude and oxygen limitations unless your aircraft is powerful and properly equipped. If it is, this is a good way to avoid a thunderstorm.

Through a Saddleback

If your plane cannot climb over the top, maybe you can pick your way through between saddlebacks. However, you may still experience hail and turbulence.

Flying Under the Base

Whenever circumnavigation is impossible, flight can sometimes be made under the base of the storm if it lies over the sea or flat, open country. Never fly under the storm in mountainous areas unless there is a definite ceiling with good visibility underneath. Quite frequently, storms in mountains have their bases below and often around peaks.

Radar and Thunderstorm Flying

Radar is, without question, one of the best instrument aids to thunderstorm flying. The regions of maximum turbulence and drafts coincide with the regions of high water content and, consequently, are within the area delineated by the radar echo. Radar is particularly useful for planning flights at higher altitudes, where there is considerably greater distance between convective centers than at low altitudes. In daylight, these open areas might be visible, but they are not visible in darkness or in the presence of enshrouding cloud layers. In such cases, indications from the radar equipment should enable you to select a flight course which will completely avoid turbulent areas. (See Chapter 18 for more detail about radar and flying.)

In the case of squall lines, or "roll" clouds, it is possible to pick a route that will take the aircraft through the line in the least time and will result in encountering the least turbulence. Even in intense squall lines, it is seldom that the cells are so close to one another that you cannot find a flight path through it without having to detour excessively from your original flight plan. If your aircraft is equipped with a radar set having a reasonably long range and a PPI scope, you may observe the structure of the squall line while still many miles away and take action to plan your penetration or circumnavigation with a minimum loss of time.

Preparing for Thunderstorm Penetration

(1) Keep in touch with the ground on the assigned *Metro* frequency. Ask for clearance to deviate from course. If radar tracking is available, use it to avoid the storm area. Know your position before you penetrate. Get mentally as well as physically prepared.

(2) Since both direction and attitude of the flight are to be maintained by reference to the gyro instruments, they should be checked carefully before entering the storm. Set your gyro horizon bar. Lower the seat if this will help you.

(3) To guard against temporary blindness from bright flashes of lightning, wear dark glasses even at night, and turn on the cockpit lights full bright. Turn off the anticollision light (to avoid vertigo).

(Some pilots close one eye during lightning to preserve night vision in the other eye.)

(4) Turn off all radio equipment rendered useless by static.

(5) Use VHF or UHF frequencies which are not significantly affected by static.

(6) The carburetor mixture should be set *full rich*. Set the props as desired.

(7) Slow down—but don't get so slow you flounder into a stall. The faster a plane is flying when it strikes an updraft or downdraft, the greater the shock will be. The operator's manual for the particular aircraft gives the correct range of speed for penetration. A speed that is about 50% above stalling is a good thumb rule.

(8) If hail is encountered, slow the aircraft to the minimum speed recommended for flights in severe turbulence.

(9) Since ice formation in the Pitot tube and carburetor is probable, turn on both Pitot heat and carburetor heat.

(10) Stow all loose gear.

(11) Turn off any electrical equipment you don't need. It may save damage from a voltage surge caused by a lightning strike.

(12) Put your fuel selector on your best tank.

Penetrating the Thunderstorm

If you can't circumnavigate, if you can't find a saddleback, if you can't go under, and if you *must* penetrate, here is the distilled experience of hundreds of pilots on how to do it—and good luck!

(1) Fasten your seat belt.

(2) Fly at the airspeed or power setting recommended in the operator's manual.

(3) Establish airspeed and power settings before entering the storm.

(4) Fly at constant altitude using your gyro. Keep a constant power setting as far as possible. Use your throttles only when high or low airspeed limits are exceeded. Don't worry about the rate-of-climb instrument or altimeter. They may fluctuate wildly. Don't change the airspeed. Erratic reading of airspeed results from vertical drafts past the Pitot tube and the clogging effects of rain and ice.

(5) Avoid all unnecessary maneuvering—to prevent adding maneuver loads to the loads already being imposed by the turbulence.

(6) Avoid use of the autopilot. The autopilot, being a constant-

altitude device, will dive the aircraft to compensate for an updraft and climb it in a downdraft. In updrafts, excessive airspeeds may be built up, and in downdrafts the airspeed may approach the stalling speed.

(7) Hold a constant heading through the storm. Wandering will only prolong the flight and increase the dangers.

(8) Once in, *don't turn back,* since a change in direction will place additional stress on the aircraft, and may prolong the exposure time. You also invite vertigo.

(9) When flying under the storm, choose an altitude one-third of the distance between the ground and the base of the clouds and fly around the heavy rainstorm areas.

(10) If you are crossing a line of thunderstorms, cross at a right angle in order to minimize the time in the turbulent area.

(11) If your airplane has a "bird-dog" radio direction finder, remember that it may be attracted to a thunderstorm more than to the station itself.

(12) Don't skim the tops of thunderstorms. Some of the worst turbulence occurs just above the tops.

16 ICE AND FLYING

GENERAL

So far in Part IV, we have covered two of the three most hazardous weather problems a pilot faces—fog and thunderstorms. The third weather problem which is just as hazardous as these two is ice. Good flight instruments and better meteorological services have minimized the danger of fog and thunderstorms so that a pilot can usually avoid them. The problem of ice can also be anticipated, but the technique of handling icing is somewhat more complex.

First of all, let us define icing as that condition which affects airframes, engines, and propellers or rotors, caused by an accumulation of supercooled water drops which form ice on these and other exposed surfaces of an aircraft and, under certain conditions, inside their power plants.

The presence of ice on an airfoil disrupts the smooth flow of air, thereby decreasing lift and increasing drag. In fact, drag may be increased by as much as 35% and lift reduced as much as 50%. Ice on the propeller causes a loss of efficiency which results in a loss of thrust. This can amount to a 19% efficiency loss. The addition of ice to the various structural parts of the aircraft also results in vibration, causing added stress on these parts. This is especially true in the case of the propeller, which is delicately balanced. Even a small amount of ice, if not distributed evenly, can cause great stress on the propeller and engine mounts. And when a prop sheds some of its ice, a momentary increase in vibration and stress occurs. Thus, the danger of added weight in combination with lift and thrust losses, plus vibration stress, become important factors in critically loaded aircraft.

Icing is not restricted to airfoils and other external structure. Reciprocating engines, carburetors, and jet engines with axial flow compressors are also affected by ice formation. Internal ice at the engine intakes, in the carburetor, or inside the engine itself, chokes the engine and reduces its power. Ice also affects canopies, Pitot-static systems, and propellers.

The total effects of aircraft icing, therefore, are:

- Loss of aerodynamic efficiency
- Loss of engine power
- Loss of proper operation of control surfaces, brakes, and landing gear
- Increased vibration and structural stress
- Loss of aircrews' outside vision
- False flight instrument indications
- Loss of radio communication

Ice can form as slowly as a half-inch per hour or as rapidly as one inch per minute!

Is there any wonder, therefore, that ice is considered one of the three most severe weather hazards a pilot faces?

TYPES OF ICE

Having discussed the several effects of ice, let us now discuss the three main types of ice—frost, glaze, rime—and their characteristics.

First of all, three conditions are necessary for formation of either rime or glaze ice (not frost, however): (1) the presence of visible moisture, (2) an outside air temperature at or below freezing, and (3) an aircraft whose temperature is at freezing or less. The U. S. Air

Force's Air Weather Services records indicate that the relative frequency of these ice types is: clear, 10%; clear-rime mixture, 17%; rime, 72%; and frost in flight, only 1%.

Frost

Frost occurs most frequently on aircraft parked outside at night, just as it does on a parked automobile. As the aircraft surface temperature, through radiational loss of heat, drops below freezing and below the saturation temperature of the air, the moisture in the air can sublime on the aircraft in the form of frost. (Frost can also form on an aircraft making a rapid descent through clear air when the aircraft surface temperature is below freezing and below the saturation temperature of the air. The formation of frost in flight is considered rare and harmless—not so with frost on an aircraft preparing to take off.)

The effect of frost is more subtle than that of ice. While the formation does not change an airfoil's aerodynamic contour, it does provide a surface of considerable roughness, with resulting increase in drag and incipient stalling of the wing.

Under no circumstances should a takeoff be attempted with a frost formation on the aircraft. A heavy coating of hard frost on the upper surfaces of the wings can cause a 5 to 10% increase in stalling speed. The increase in drag will not significantly affect the initial acceleration during takeoff; but the takeoff speed is generally 5 to 25% greater than the stall speed. It is possible, therefore, that with a frost formation, an aircraft cannot become airborne at the specified takeoff speed because of premature stalling. If the aircraft does become airborne at the specified takeoff speed, it can experience incipient or complete stall in turbulence, gusts, or turning flight because of an insufficient margin of airspeed above stalling speed.

Just as there is no such thing as a *little* pregnancy, there is no such thing as a *little* frost. *Any* frost on an aircraft can be dangerous.

Rime Ice

Rime ice (see Figure 16–1) is a rough, milky, opaque deposit which is formed by the instantaneous freezing of *small* supercooled drops as they strike the aircraft. They maintain their spherical shape as they freeze, trapping air between them, making a porous and brittle ice formation Since it is formed by small drops, rime builds at a slower rate than glaze ice. Its porous formation is less cohesive than the

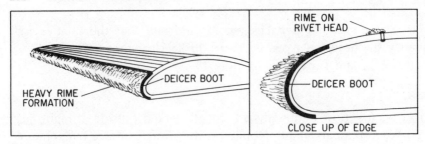

Figure 16–1 Rime Ice on an Airfoil.

sheet-like glaze formation. Rime ice is a more streamlined formation than glaze, thus causing less distortion of the airflow. Fast freezing to form rime ice can take place when the temperature is anywhere from 0°C. to −40°C., but is most probable between −10°C. and −20°C.

Rime ice is three or four times more common than glaze ice.

Clear Ice

Clear ice (see Figure 16–2) is the most serious of the three ice types because it adheres so tenaciously to an aircraft. It is formed by the relatively slow freezing of *large*, supercooled liquid-water droplets, which have a tendency to spread out and assume the shape of the surface on which they freeze. As a result of the spreading of this supercooled water and its slow freezing, very few air bubbles are trapped within the ice, as in rime ice, which accounts for its clearness. Although clear ice is expected mostly with temperatures between 0°C. and −10°C., it can occur with temperatures as cold as −25°C.

Clear ice can be expected in clouds with appropriate temperatures where vertical currents can support large drops. These conditions can be expected when the following are present:

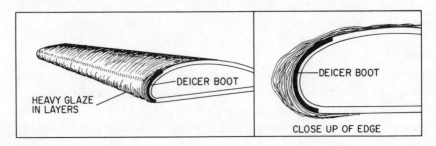

Figure 16–2 Clear Ice on an Airfoil.

(1) Convective action from surface heating
(2) Vertical convection forced by cold fronts
(3) Active up-glide of warm air over the warm front
(4) Up-slope lifting

ICE FORMS

There are several forms of ice which an aviator will encounter—snow, sleet, hail, propeller ice, rotor ice (in helicopters), windshield and canopy ice, Pitot tube ice, antennae ice and structural ice.

Snow

When condensation takes place at temperatures below freezing, water vapor changes directly into minute ice crystals. A number of these crystals unite to form a single snow flake. Dry snow does not lead to the formation of structural ice, but wet snow does.

Sleet

When raindrops fall through the air that is below freezing, they may freeze to form sleet. This form of ice is a cold-weather phenomenon. It does not bring about the formation of structural ice, except when it is mixed with supercooled water. Supercooled water can be either in small droplets (which forms *rime* ice) or large droplets (which forms *clear* or *glaze* ice). In either case, supercooled water vapor is relatively unnatural, but it is still frequently encountered in flight.

Hail

This form of ice develops in highly turbulent thunderstorms. Water drops, which are carried upward by vertical currents, freeze into ice pellets, start falling, accumulate a ring of water, and are carried upward again. The newly added water freezes. A repetition of this process increases the size of the hailstone. Hail is a warm-weather phenomenon and can be produced only by strong vertical currents. It does not lead to the formation of structural ice, but, in some cases, it causes physical damage to the aircraft.

Propeller Ice

Ice deposited on the propeller can reduce its aerodynamic efficiency, produce serious vibration and stress, and cause increased fuel consumption. The lower the RPM, the more favorable the conditions for ice formation.

Antennae Ice

On the small-diameter wires of a radio antenna, a tacan or omni antennae, ice can form faster and sooner than it forms on the wings or the propeller. Such formation can be sufficiently heavy to cause the loss of the antenna itself. Some pilots attempt to remove antennae ice by bridging the insulators, thus grounding the antennae to the aircraft frame and causing some heating.

Carburetor Ice

Carburetor ice (see Figure 16–3) constitutes a problem different from that of structural ice; it can form when it is impossible for structural ice to form. For this reason, it is treacherous. If the humidity is high, carburetor ice may occur with air temperatures as high as 25°C. The cooling effects of the reduced pressure produced by flow of air in the Venturi tube and the evaporation of gasoline are sufficient to cause ice to form.

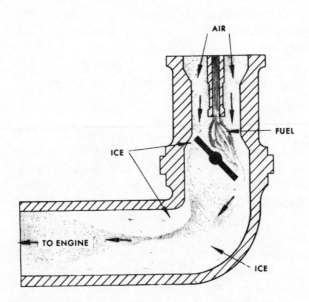

Figure 16–3 Carburetor Ice.

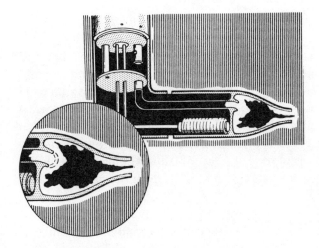

Figure 16–4   Pitot-tube Ice.

A pilot should remember that the weather forecasting job is to forecast weather conditions, including the forecasting of airframe icing. It is *not* the meteorologist's responsibility to forecast *carburetor* or *engine* ice. This is the *pilot's responsibility*.

Pitot Tube Ice

Ice in the Pitot tube (see Figure 16–4) reduces its size and changes the flow of air in and around it, making the airspeed instrument useless or unreliable. Turn indicator, rate of climb, and other instruments depending on Pitot air are also affected by icing. Pitot icing is usually easy to prevent with the use of Pitot heat. If it *is* clogged by ice, the pilot must fly attitude and use proper power settings and *not* depend on an unreliable airspeed indicator.

Helicopter Ice

The formation of ice on a helicopter is not as critical as on a fixed-wing aircraft. This is not to say, however, that a helicopter is invulnerable to ice. Even a $\frac{3}{16}$-in. coating of ice on a helo's main rotors can prevent flight. But a helicopter does have some advantages over a regular airplane: The centrifugal force of rotation, the bending and flapping of the blades during rotation, the slow rotational speed of the blades near the rotor head, the fast rotational speed near the blade tips, and the slow flight of the helo itself—all tend to reduce ice accumulation.

Helicopter icing occurs in two principal areas—the main rotor and the tail rotor. Ice formation on either rotor can produce critical vibration and loss of efficiency or control. Although the slow forward speed of the helicopter reduces ice build-up on the fuselage, the rotational speed of main and tail rotor blades produces a rapid build-up rate on certain surface areas. A critical icing hazard can therefore form rapidly on the center two-thirds of the main rotor blades. The uneven build-up or shedding of this ice formation can also produce severe rotor vibration.

(1) *Main Rotor Head Assembly* Ice accumulation on the swashplates, push-pull rods, bell cranks, hinges, scissors assemblies, and other mechanisms of the main rotor head assembly interferes with collective pitch and cyclic control.

(2) *Tail Rotor* Ice accumulation on either the tail rotor head assembly or tail rotor blades produces the same hazards as those associated with the main rotor. The centrifugal force of rotation and the blade angle of incidence relative to the clouds help to reduce ice build-up on the tail rotor blades, but the shedding of ice from the blades may damage the fuselage or the main rotor blades.

Windshield and Canopy Ice

During instrument flight, a pilot can tolerate windshield or canopy ice, but at time of landing he must have good forward visibility. Various deicing systems have been designed to do this. Some have heating wires embedded in the windshield plexiglass. Other aircraft have a defroster system of hot air like that of an automobile. Some pilots also carry a putty knife, and are able to reach outside the cockpit before an approach and scrape off the accumulated ice.

INTENSITY OF ICE

In 1964, standard designations for icing intensities were established. The intensity definitions are based on meteorological and operational criteria. The operational definitions are based upon the rate at which ice forms on a small probe (or projection from an airframe).

Trace of Icing

Accumulation of one-half inch of ice on a small probe per 80 miles flown. The presence of ice on the airframe is perceptible but the rate of build-up is nearly balanced by the rate of sublimation. Therefore, this is not a hazard unless encountered for an extended period of time. The use of deicing equipment is normally unnecessary.

Light Icing

Accumulation of one-half inch of ice on a small probe per 40 miles flown. The rate of build-up is sufficient to create a hazard if flight is prolonged in these conditions but is insufficient to make diversionary action necessary. Occasional use of deicing equipment may be necessary.

Moderate Icing

Accumulation of one-half inch of ice on a small probe per 20 miles flown. On the airframe, the rate of build-up is excessive, making even short encounters under these conditions hazardous. Immediate diversion is necessary or use of deicing equipment is mandatory.

Heavy Icing

Accumulation of one-half inch of ice on a small probe per 10 miles or less. Under these conditions, deicing equipment fails to reduce or control the hazard, and diversionary action is mandatory.

These 1964 standards were updated in 1968 and the Airframe Icing Reporting Table (Table 16–1) was issued. (This table was developed and approved by the Subcommittee on Aviation Meteorological Services of the Interdepartmental Committee for Meteorological Services.)

Pilots are expected to know the above criteria so that they can report ice conditions accurately to the meteorologist after landing or make pilot reports (PIREPs) during flight. (See Chapter 11, page 169.)

TYPES OF WEATHER CONDITIONS WHERE ICE IS FOUND

The two weather conditions conducive to icing occur:
- (1) Under a frontal inversion
- (2) Whenever liquid water (fog, cloud, mist, or rain) is found at subfreezing temperatures

Table 16–1   Airframe Icing Reporting Table

INTENSITY	ICE ACCUMULATION	PILOT REPORT
Trace	Ice becomes perceptible. Rate of accumulation slightly greater than rate of sublimation. It is not hazardous even though deicing/anti-icing equipment is not utilized, unless encountered for an extended period of time—over one hour.	Acft Ident., Location, Time, (GMT) Intensity of Type,[a] Altitude/FL, Aircraft Type, IAS
Light	The rate of accumulation may create a problem if flight is prolonged in this environment (over one hour). Occasional use of deicing/anti-icing equipment removes/prevents accumulation. It does not present a problem if the deicing/anti-icing equipment is used.	Example: Holding at Westminister VOR, 1232Z
Moderate	The rate of accumulation is such that even short encounters become potentially hazardous and use of deicing/anti-icing equipment or diversion is necessary.	Light Rime Icing, altitude 6000, Jetstar IAS 200 kts
Severe	The rate of accumulation is such that deicing/anti-icing equipment fails to reduce or control the hazard. Immediate diversion is necessary.	

[a] Rime Ice: Rough, milky, opaque ice formed by the instantaneous freezing of small supercooled water droplets.
   Clear Ice: A glossy, clear or translucent ice formed by the relatively slow freezing of large supercooled water droplets.

Under a Frontal Inversion

When moist, warm air is forced to rise over a colder air mass, a frontal inversion will exist. Below it, icing dangers are frequently encountered. In being forced aloft, the air may experience sufficient cooling to produce saturation and precipitation. The raindrops falling into the cold air may freeze upon contact with the airplane if the temperature is at or below freezing.

The cloud system above the front may be stratified or cumuliform. The drops may be small or large and the icing may be rime, glaze, or both. Often the glaze is mixed with sleet and snow. A pilot finding himself in an icing zone under such an inversion should climb to the warm air above, if he knows just where the front lies and can be sure of finding the warm air in a short time. Otherwise he should turn around.

In Stratiform Clouds

Stratified clouds are always indicative of stable air. Either tiny fog particles, small drops of drizzle and rain, or ice crystals are present in them. Ice crystals offer no icing problems, since they do not stick to the airplane upon impact. The supercooled liquid water in the form of fog or drizzle will be frozen immediately into rime upon contact with the plane.

Glaze ice may form in the rain zones of these clouds, and often rime and glaze will form together. Flight should be either *under* these zones (where the temperature is above freezing) or *above* them (where there is no liquid water).

Since icing occurs in stratiform clouds, aircraft should be flown under the icing zones where the temperature is above freezing, or above the zones where only ice crystals are present.

In Cumuliform Clouds

Cumuliform clouds are the result of unstable air containing great quantities of water vapor. It may be lifted by surface heating, or by frontal and up-slope influence, to the level of free convection, where its great energy is released for further vertical development.

When the temperature in cumuliform clouds is reduced below the freezing point, icing dangers exist. The vigorous vertical turbulence is able to support relatively large liquid drops, which can remain liquid at low temperatures. Upon impact with the plane, these large drops spread out, forming glaze ice. Glaze ice sticks tenaciously to the plane. Since it can accumulate rapidly, it becomes a serious hazard in a short time.

Along Warm Fronts

The stratified cloud system associated with the up-glide of warm, moist air over a warm front often reaches dangerous subzero temperatures, where rime or glaze icing will occur. If the warm air mass is conditionally unstable, cumuliform clouds may form. These clouds develop conditions favorable to the formation of glaze or a combination of glaze and rime.

Figure 16–5 shows a typical warm-front structure, with the most probable icing zones and a possible flight path to avoid icing. The cloud system of a warm front is extensive. A flight through such a system is a long one; therefore, icing dangers are increased.

Along Cold Fronts

Cumuliform clouds are associated with cold fronts. Glaze ice, therefore, can be expected in them. The cloud systems are usually relatively

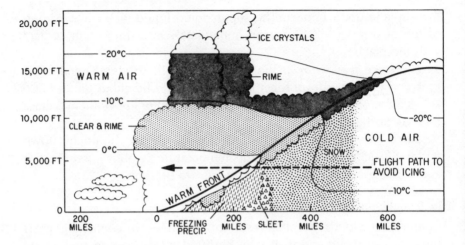

Figure 16-5 Warm-front Icing Zones.

narrow as compared with those of the warm front, and the time required to fly through them is usually comparatively short. However, due to the heavy precipitation, icing can be rapid and extremely dangerous, even though it is of short duration. A possible flight path to avoid extended icing through the cloud system of a cold front is given in Figure 16-6.

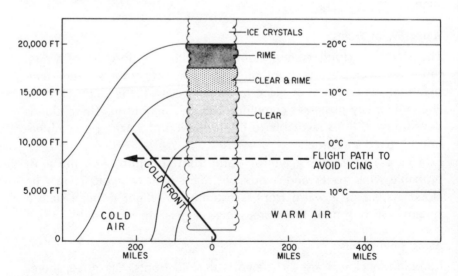

Figure 16-6 Cold-front Icing Zones.

About 85% of the observed icing of aircraft in the United States is associated with frontal systems.

Along Occluded Fronts

Occluded fronts present less of an icing hazard because, in these situations, precipitation has been occurring for some time. Consequently, there is less liquid water available for ice formation and the rate of deposit is lower. The cloud cover in an occlusion, however, is generally extensive, and the time spent in these clouds will be great. Figure 16–7 shows icing zones along an occluded front.

Over Mountains

The lifting of unstable air over mountain ranges is one of the most serious ice-producing processes experienced in the United States. Such air moving northward and eastward over the Appalachian Mountains is often cooled to subzero temperatures. An icing hazard exists for all air traffic that must travel through this air. Similarly, polar air in winter, approaching the west coast of the United States, contains considerable moisture in its lower levels. As it is forced aloft by the successive mountain ranges encountered in its eastward movement, severe icing zones develop.

Figure 16–8 illustrates a section across parallel mountain ridges, showing regions of severe icing. These icing zones appear some distance above the crest of the mountains and may continue to considerable heights. The most severe icing will take place above the crests

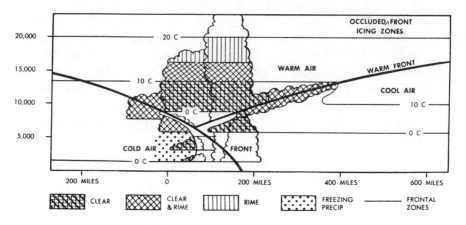

Figure 16–7 Occluded-front Icing Zones.

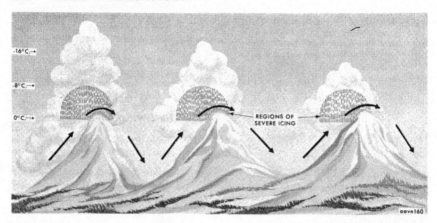

Figure 16–8   Icing over Mountains.

and to the windward side of the ridges. Usually the zone extends about 4000 ft above the tops of the mountains. In the case of unstable air, it may be higher.

The movement of a front across a mountain brings together two important factors that aid in the formation of icing zones. A study of icing in the western United States has shown that almost all the ice cases occurred where the air was blowing over a mountain slope or up a frontal surface, or both.

ICING HAZARDS BEFORE FLIGHT

The danger of accumulated frost on parked aircraft has already been mentioned. Snow or frozen rain on an aircraft is equally dangerous. If not adequately removed, they can cause considerable danger to the flight. Therefore, it is common practice to place an aircraft in a hangar until the ice melts, or to completely clean wing and tail surfaces prior to flight. If the aircraft is hangared for deicing, keep in mind that when it comes out of the hangar there will probably still be some water left on the surfaces and it will ice again if freezing temperatures exist.

Another serious condition arises from the presence of water or mud on ramps and runways during freezing temperatures. Water blown or splashed on the aircraft forms an ice cover on the under side of flaps, control surfaces, and landing gear. Mud and water freezing in the brakes and in the landing gear mechanism can cause braking failure or improper operation. Be alert and conduct ground operation accordingly. Furthermore, after taking off from a runway covered with snow or water near freezing, a pilot should exercise his brakes and operate the landing gear before its final retraction into the wells.

ICE INDICATIONS IN FLIGHT

Ice on the wings or tail is an indication that there will also be induction-system ice in the power plant. (However, induction-system ice can form under conditions which will not produce structural ice, so absence of structural ice is *not* an indication that there will be *no* carburetor ice.)

One of the best indicators of induction icing is a loss of manifold pressure with constant altitude and constant power setting on an aircraft with a constant-speed propeller. If the engine has a manifold-pressure regulator, there may be no indication of ice until the throttle is wide open.

Ice on fuel-metering parts may be indicated by the flow meter, surging, engine roughness and backfiring, or an inability to operate at reduced power.

Ice on the throttle valve or in the vicinity of it may cause engine roughness. Ice on the throttle may also cause the throttle to stick.

AIDS TO COMBATING ICE IN FLIGHT

There are three common aids for minimizing ice formation while in flight: (1) mechanical aids, (2) deicing fluids, and (3) heat.

Mechanical Aids

The leading edges of wing and tail surfaces of some aircraft are equipped with rubber skins, or boots, that normally assume the contour of the airfoil. Compressed air is cycled through ducts in these rubber boots causing them to swell and change shape. The stress produced by the pulsating boot cracks the ice and the air stream peels it off.

Anti-Icing Fluids

These are used on rotating surfaces where centrifugal force spreads the fluid evenly. Such fluids are effective agents because the fluid helps prevent ice from adhering to the surfaces and the centrifugal force throws off the ice as it forms.

Heat

Heat application capability to wing, props, and tail surfaces is installed in most newer aircraft. The icing areas are heated by electrical means or by hot air which is piped from the engine manifold or jet engine compressor. This process has frequently given rise to the name *hot wing*. Few light aircraft, however, are equipped with hot wings.

There are two basic ways for combating carburetor ice: (1) use of alternate air, or (2) preheating.

Alternate air   The design principle of the alternate air system is that a relatively low temperature rise will prevent the formation of ice with injection-type carburetors if free moisture can be excluded from the air intake.

Such a temperature rise will, in turn, permit continued and unattended operation in alternate air without the risk of detonation from excessive carburetor air temperatures. On the other hand, the low temperature rise is not considered sufficient to remove ice that has accumulated to an appreciable degree. The shift to the alternate air system must, therefore, be made either *before* icing conditions are actually encountered or immediately thereafter.

To achieve the desired result, the alternate air system is designed with a two-position door in the induction system. When the door is placed in the alternate position, the main air intake scoop is sealed off to prevent the entrance of free moisture, and warm air is drawn into the induction system either from the engine accessory compartment or from behind the engine cylinders.

The alternate air position should be used when flying in any conditions conducive to the formation of induction ice. This means operation in rain, snow, sleet, low-temperature clouds, and in cool air of relatively high humidity as encountered near clouds.

Preheating   The design principle of preheating is that when cool, moist air entering the carburetor is mixed with air of high temperature from an exhaust muffler or preheater, the resultant mixture has a temperature high enough to prevent the formation of ice.

The heat available in this system must be sufficient to evaporate at least part of the free moisture entering the induction system. Sufficient temperature rise is generally provided to melt ice formations which might occur before the application of heat.

Preheating should be used when flying in any conditions conducive to the formation of ice. While it is possible to wait until carburetor icing is actually encountered before the application of heat, the practice is not recommended. Many pilots follow this practice, mistakenly believing that engine efficiency is lowered by the use of carburetor heat. As a general rule, for conditions where it is necessary to use heat intermittently to remove ice formations, the constant use of heat to prevent icing will provide improved engine performance. By waiting too long before applying preheating, engine power may be reduced to a point at which insufficient heat is available for recovery.

Preheat should always be maintained for approach or landing under carburetor icing conditions. Most float-type carburetors are very susceptible to icing. Even in clear air at nearly closed throttle positions, the upper icing limits are as high as 25°C. carburetor air temperature in some types. Therefore, for aircraft equipped with float-type carburetors, preheat should be used in most instances during letdown, approach, and landing.

Preheat should be used for takeoff under carburetor icing conditions. Preheat should be used for takeoff in outside air temperatures of 0°C. or lower and is strongly recommended at temperatures of −20°C. or lower.

FLIGHT TECHNIQUES IN ICING CONDITIONS

If the aircraft is not equipped with wing deicers and ice is encountered, turn back to the nearest airport and use power to land, keeping airspeed well above normal landing speed.

If the aircraft is equipped with leading edge boot-type deicers, don't use them too quickly. Allow ice ⅛ to ¼ in. thick to form, then crack it off by inflating the deicers. After initial use, use the deicers periodically as ice forms to prevent the ice from creating a pocket in which the boot expands.

Planes with *hot wings* should be prepared *prior* to encountering icing conditions. Turn on the heat to prevent ice from forming.

If you have ice on your wings when landing, use more power to keep the airspeed above the stalling point. Land at the nearest airport with power and, if possible, avoid turns at low altitude.

If the aircraft is not equipped with slinger rings or other propeller anti-icing equipment, a change in speed of the engine or a change in propeller pitch may loosen the ice on the blades of the propeller. If the aircraft is equipped with slinger rings, the fluid should be turned on *before* encountering the icing condition and flow should be maintained until the aircraft is in the clear and out of the icing zone.

Keep the controls moving slightly during flight through an icing zone. This prevents freezing of control surface hinges. Planes equipped with windshield defrosters should have them turned on *before* entering the icing zone.

The following precautions should be kept in mind:

(1) Do not continue flight into a region of known icing conditions.

(2) Do not fly through rain showers or wet snow when temperature at flight level is near freezing.

(3) Do not fly parallel to a front under icing conditions.

(4) Do not fly into clouds at a low altitude above the crests of ridges or mountains. Fly 4000 or 5000 ft above ridges when flying on instruments through clouds at indicated free air temperatures greater than plus 2°C.

(5) Do not fly into cumulus clouds at low temperatures. Heavy glaze may be encountered.

(6) Do not make steep turns with ice on the airplane.

(7) Do not land with power off when ice has formed on the wings and other exposed surfaces of the plane.

(8) Do not try to climb too fast when ice has formed on the plane, since stalling speed is higher than normal.

(9) Do not forget, when flying under icing conditions, that fuel consumption is greater, due to increased drag and the additional power required.

(10) Do not attempt a cross-country flight in fall or winter without first consulting the nearest weather office to obtain a forecast as to expected icing conditions.

(11) Do not remain in icing conditions too long when landing. (See Figure 16–9.)

(12) Change altitude (climb or descend in layer-cloud icing) and vary course as appropriate to avoid cumulus-cloud icing.

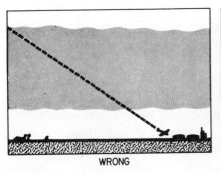 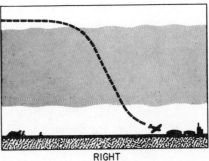

Figure 16–9 Landing Technique in Icing Conditions.

(13) If you have found it necessary to use alternate air in flight, keep it on for the landing. (However, don't take off in alternate air unless there are actual icing conditions present—and if you do, remember you won't get full power.)

(14) Climbs and descents through icing conditions should be made at as rapid a rate and at a speed as slow as practicable.

Remember these points about aircraft icing:

(1) Icing hazards above clouds are not great.

(2) Ice clouds are not hazards, since the ice will not adhere to the plane.

(3) Icing is severe in frontal zones and in up-slope conditions over mountains and fronts.

(4) Icing can be expected in rain or drizzle below a cloud when the free air temperature gauge reads less than +2°C.

(5) Speed of an aircraft affects the rate of ice accumulation. The faster an aircraft moves through an area of supercooled drops, the more moisture it encounters and the faster will be the ice accumulation.

CHECKOFF LIST FOR COLD-WEATHER OPERATIONS

Following is a winter checkoff list that will help reduce hazards of cold weather flying:

(1) Check weather carefully; ask the pilot who just came through.

(2) Check NOTAMs.

(3) Remove frost and snow before takeoff.

(4) Check controls for restriction of movement.

(5) Taxi slowly. Use brakes with caution. Avoid pools of runway water when it is near freezing.

(6) After runup in fog or rain, check wing and empennage for ice in propeller wash area.

(7) Wear sunglasses if snow glare is bad.

(8) Avoid taking off in slush or wet snow, if possible. Exercise the brakes and landing gear after takeoff before final retraction.

(9) Be alert for snowbanks during takeoff and landing.

(10) Use Pitot heater when flying in rain, snow, clouds, or known icing zones.

(11) When flying in freezing-rain conditions, climb into the clouds where the temperatures will be above freezing (unless the temperature at a lower altitude is *known* to be high enough to prevent ice).

(12) Report all in-flight weather hazards.

(13) If icing can't be avoided, choose the altitude of least icing. (Glaze ice is common in cumulus clouds; rime ice is common in stratiform clouds.)

(14) Use carburetor preheat to prevent ice formation; do not wait until an icing condition exists. Watch the carburetor air temperature, especially between $-5°C.$ and $+10°C.$ Use full carburetor heat to clear it of ice.

(15) Watch airspeed. Stalling speed increases with airframe icing.

(16) Check wing deicers; use them properly. Do not land with deicers on since they act as airflow spoilers.

(17) Avoid making steep turns if the aircraft is heavily coated with ice.

(18) Avoid making three-point landings when "iced up." Fly in with power.

(19) Before starting a landing approach, slowly move throttle back and forth to make sure the carburetor butterfly valve is free of ice.

(20) Maintain carburetor heat whenever carburetor icing is likely.

(21) Before takeoff, insure that anti-icing and deicing equipment is in operating condition.

17 TURBULENCE AND FLYING

TURBULENCE

One of the most unpredictable weather phenomena of special significance to pilots is turbulence. Turbulence can be so violent as to throw you out of your seat, bump your head on the canopy, or cause airsickness. In the parlance of the pilot, turbulence is frequently described as "hitting an air pocket" or "downdraft" or "gusts."

Turbulence is caused by random fluctuations of wind flow which are instantaneous and irregular. There are four degrees of turbulence which you should know: (1) light, (2) moderate, (3) severe, and (4) extreme. Here are the explanations of these four degrees:

Light turbulence is a condition during which occupants may be required to use seat belts, but objects in the aircraft remain at rest. The airspeed may fluctuate from 5 to 15 knots. Usually found at low levels over rough terrain with surface wind speed less than 25 knots.

Moderate turbulence is a condition in which occupants require seat belts and occasionally are thrown against the seat belt. Unsecured objects in the aircraft move about. The airspeed may fluctuate 15 to 25 knots. Usually found in and around dissipating thunderstorms, or within their cirrus tops; near rough terrain with surface winds about 25 knots; in the jet stream and in towering cumulus.

Severe turbulence is a condition in which aircraft may be momentarily out of control. Occupants are thrown violently against the seat belt and back into their seat. Objects not secured are tossed about. Airspeed fluctuates more than 25 knots. Usually found in and around *severe* thunderstorms, in jet streams, and on the lee side of mountains where the wind is blowing across them at 20 or more knots.

Extreme turbulence is a rarely encountered condition in which the aircraft is violently tossed about and is practically impossible to control. May cause structural damage. Airspeed will make rapid fluctuations in excess of 25 knots. Extreme turbulence is found on the lee side of mountains across which winds of 50 knots or more are blowing, or in the biggest, most mature "thunder-bumpers."

There are four causes of turbulence:

(1) *Thermal* (or *convective*) *turbulence* is caused by localized vertical convective currents due to surface heating or unstable lapse

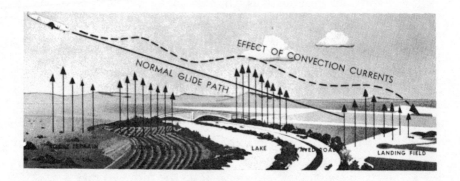

Figure 17-1 Thermal turbulence ("bumpy air").

Figure 17–2 Mechanical Turbulence.

rates, and cold air moving over warmer ground. More explicitly, warm air rises and cold air descends, and at the boundary of the two, a pilot encounters "bumpy air." Figure 17–1 is an example.

(2) *Mechanical turbulence* results from the friction of wind flowing over irregular terrain or man-made obstacles. Figure 17–2 is an example.

(3) *Frontal turbulence* results from the local lifting of warm air by cold air masses, or the abrupt wind shift (shear) associated with most cold fronts.

(4) *Large-scale wind shear* is a marked gradient in wind speed and/or direction due to general variations in the temperature and pressure fields aloft.

Two or more of the above causative factors often work together.

In addition to these four, turbulence is produced by "man-made" phenomena, such as the wake of an aircraft.

Pilots can avoid thermal (convective) turbulence by flying above the levels reached by convective currents. You can usually tell you are above this type of turbulence by staying above the tops of cumulus clouds or the haze layer. (See Figure 17–3.)

If you are a glider pilot, thermal turbulence can give you the altitude lift you are seeking. (See Chapter 20.)

Pilots encounter mechanical turbulence when the air near the surface of the earth flows over obstructions, such as irregular terrain

Figure 17-3 Stay Above the Tops of Cumulus Clouds to Avoid Turbulence.

(bluffs, hills, and mountains) and buildings. In this event, the normal horizontal wind flow is disturbed and transformed into a complicated pattern of eddies and other irregular air movements. For a light plane particularly, such turbulence can make landings and takeoffs interesting if not hazardous. (See Figure 17-4.)

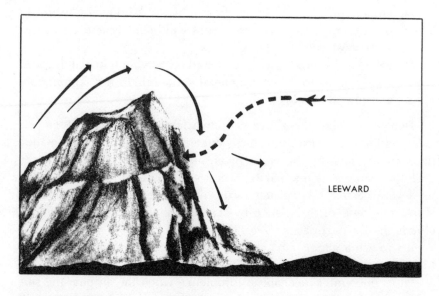

LEEWARD

Figure 17-4 Mountain-wave Turbulence.

The strength and magnitude of mechanical turbulence depend upon the speed of the wind, the roughness of the terrain (or nature of the obstruction), and the stability of the air. Stability seems to be the most important factor in determining the strength and vertical extent of the mechanical turbulence.

When strong winds blow approximately perpendicular to a mountain range, the resulting turbulence may be quite severe. This condition is referred to as *standing waves* or *mountain waves,* and may or may not be accompanied by turbulence. Pilots, especially glider pilots, have reported that the flow in these waves is often remarkably smooth. Others have reported severe turbulence.

One of the most dangerous features of the mountain wave is the turbulence which produces downdrafts just to the lee side of the mountain peaks and to the lee side of the roll clouds which are formed. (See Figure 17–4.)

RULES FOR FLYING THE MOUNTAIN WAVE

There are six rules for flight over mountain ranges where mountain waves exist:

(1) If possible, the pilot should fly around the area when wave conditions exist. If this is not feasible, fly at a level which is at least 50% higher than the height of the mountain range. Be cautious to attain or maintain the minimum safe altitude (terrain elevation plus 50%) during climb-out to cruising altitude, and descents for landings.

(2) Avoid roll clouds, since they are the areas with the most intense turbulence.

(3) Avoid the strong downdrafts on the lee side of mountains. There are many instances of aircraft accidents caused by these downdrafts.

(4) Avoid high lenticular (lens-shaped) clouds, particularly if their edges are ragged.

(5) Do not place too much confidence in pressure altimeter readings near mountain peaks. They may indicate altitudes which are more than 1000 ft higher than the actual altitude.

(6) Penetrate turbulent areas at airspeeds recommended for your aircraft.

Frontal turbulence is caused by the lifting of warm air by a frontal surface leading to instability and/or the mixing or shear between the warm and cold air masses. The vertical currents in the warm air are strongest when the warm air is moist and unstable. The most severe cases of frontal turbulence are generally associated with fast-moving cold fronts. In these cases, mixing between the two air masses as well as the differences in wind speed and/or direction (wind shear) add to the intensity of the turbulence.

CLEAR-AIR TURBULENCE

One of the major problems of high-performance aircraft is clear-air turbulence (called CAT). CAT is discussed at greater length in Chapters 10 and 19.

Turbulent flight conditions are frequently encountered in the vicinity of the jet stream where large wind shears in the horizontal and vertical are often found. Since this type of turbulence may occur in perfectly clear air without any visual warning in the form of clouds, it is often referred to as clear-air turbulence. This does not mean that all clear-air turbulence occurs in clear air—only about 75% does; in fact, most frequently, CAT is associated with jet streams and mountain waves.

18 RADAR AND FLYING THE WEATHER

One of the best aids available to the meteorologist for weather forecasting is radar. It is also a great aid to pilots who are lucky enough to have one in their airplane.

The first use of the principle of radar occurred in England in the 1930's when Dr. Watson-Watt observed that radio signals were being reflected from clouds and thunderstorms. From this observation, radar developed to the point where, just before World War II, radio energy was being directed in a rotating beam and bounced off an airplane in flight. This development was a major factor for victory in the Battle of Britain in the 1940's.

Radar is a contraction of the words "radio detection and ranging." Electromagnetic energy is directed in short, sharp bursts toward a target—a thunderstorm or an aircraft, for example—and some of this energy is bounced back toward the source. This small amount of reflected energy is picked up by a receiving antenna and recorded as an echo or "blip." The time taken for the energy to travel from the source to the target and back can be measured in microseconds, but it must be done with great accuracy, since the energy is traveling at the speed of light.

From this principle, modern radar, both fixed, land-based and airborne radars, has developed the capability to detect and track severe weather phenomena.

There are several types of ground-based radar specifically designed for weather detection. These radars usually have a PPI (Plan Position Indicator scope) and an RHI (Range Height Indicator) which can measure the location, height, and intensity of storms up to a range of about 250 miles.

Any weather which contains significant precipitation can be detected on these radars. Thus, thunderstorms, hurricanes, cold fronts, and squall lines can be located and tracked with precision and their intensity measured.

To a pilot, therefore, ground-based weather radar is a great aid to safe flying for it permits the local meteorologist to improve his forecast and to help you avoid severe weather. In addition, weather radar is now standard on most large aircraft and is increasingly seen on light and private aircraft and is most useful en route.

WEATHER INDICATIONS ON RADAR

The radar picture of a thunderstorm is almost always a bright and sharp "blip" because of the multitudinous raindrop targets in the storm. Snow and ice are poorer radar reflectors than are raindrops, but they do reflect radar energy. A cold front with little weather in it will hardly show up on radar, but an active cold front and particularly a squall line will show up quite vividly on the scope. Hurricanes and typhoons also show up very well.

The brightness of the reflection on a radarscope is a direct indication of the size and number of water droplets, ice particles, and/or precipitation in the atmosphere. For example, on a standard weather-detection radarscope, many or large raindrops or hail will produce strong "blips"; fewer or smaller liquid drops and snowflakes will produce weaker echoes; and some types of clouds, fair-weather cumulus for example, will produce no reflection at all. Radar can measure rainfall rate, but not with great accuracy.

Therefore, radar shows only that portion of a storm which contains water droplets or ice crystals of sufficient size and/or number to produce radar reflections or echoes on the scope.

Once a storm area has been detected by radar, the movement and changes in size and intensity can readily be determined by making successive observations.

Radar easily detects the more intense portions of the storms enabling the pilot to fly around such areas. The intensity of the echoes, sharpness of the edges, and large size of individual storms are all indications of severity. In addition, any irregularities in the shape of the storm, such as protuberances or V-shaped notches, are often associated with especially dangerous regions, such as hail shafts.

A ground-based radar with a range height indicator (RHI) can observe the height of the storm, which is another indication of storm intensity. When this information is coupled with the rate at which the storm is building, a good indication of the degree of turbulence is obtained.

To assist in locating hurricanes, typhoons, and severe-weather storms, and to forecast them, the U.S. Weather Bureau has approximately 100 storm-detection radar sets installed along the coasts of southern and eastern United States and in the central plains where such weather is encountered.

Airborne weather radar is of great value to pilots in locating and avoiding turbulence. The intensity of precipitation is a primary indication of the amount of turbulence within a storm because strong drafts and gusts are necessary to support water drops of significant size and quantity. Moreover, since the relative intensity of precipitation can be detected by radar, a pilot can select a comparatively safe and smooth flight path in thunderstorm areas by avoiding the localized areas of heavy precipitation indicated on the radarscope.

RADAR REPORTS (RAREPS)

The surface weather radar sets now in operation are used to identify and track storms over most of the United States east of the Rockies. As has been described in Chapter 11, every hour, the stations which observe significant radar echoes transmit the information on the weather teletypewriter circuits. These transmissions are known as RAREPs. All stations, and especially those which are not equipped with radar sets, receive much valuable information from RAREPs. An example of a RAREP is as follows:

NCENTL STATES
BRKN LN STG 20 WIDE 15 SE DCA 20 NW RIC TOPS 420
WITH VRY STG CELL 45 W RIC CELLS—40.

Translated, this RAREP reports a broken storm line of strong intensity, 20 miles wide, from 15 miles southeast of Washington, D.C. (DCA) to 20 miles northwest of Richmond, Virginia (RIC). The cloud tops extend to 42,000 ft (MSL). A very strong cell is located 45 miles west of Richmond. The cells are moving east-northeast at 40 knots.

TIPS ON USING RADAR

Radar is a great aid in aircraft which have them, but it has its limitations and a pilot must know what they are. Here are a few good tips on using radar in the air:

(1) Remember that radar can't distinguish between different kinds of objects or landmarks if they provide equally good reflecting surfaces. Don't expect a map-like presentation over land.

(2) Don't count on radar to show you mountains that you might run into or navigate by unless you have some sort of height-finder arrangement.

(3) Remember that while radar can help you fly through weather, weather can damage your radar. Hail and sleet can pit and erode the plastic radome. Since the detection or forecast of hail or sleet is difficult, the best solution is to avoid all severe precipitation if you possibly can.

(4) Use radar to avoid weather. The heaviest precipitation will show up on your scope; lighter precipitation will show up as a sort of gray. The heaviest turbulence is not necessarily found in the heaviest storms; it is usually encountered at the edges of storms where there is a sharp change from heavy to light precipitation in a small space. Such a gradient will show up as a sharp drop from bright to black on the scope.

19 HIGH-ALTITUDE WEATHER FLYING

When the first edition of this book appeared in 1941, the phenomena of high-altitude flight were scarcely known. Private aircraft rarely flew above 10,000 ft, and indeed, few of them had the power to do so. Even military and commercial aircraft normally flew at altitudes now called "medium" flight altitudes, and there was scant meteorological information on such subjects as jet streams, clear-air turbulence and contrails.

All this has been changed in recent years as twin-engine private aircraft, the small jets, the light turbo-prop aircraft, and small executive aircraft operate routinely at high altitudes—above flight level 200. In the years ahead, as aircraft and their power plants continue to improve, high-altitude flight will become even more commonplace. It is necessary, therefore, that pilots operating at such altitudes understand the problems of the jet stream, clear-air turbulence, contrails, haze layers, and canopy static which they may encounter.

THE JET STREAM

In Chapter 6, there was a full discussion of wind and atmospheric circulation, and mention was made therein about jet streams, one of the distinctive problems of high-altitude flight.

Jet streams are very high-speed winds—50 knots up to 300 knots—which usually flow west to east in narrow bands in the middle latitudes between 20,000 and 40,000 ft.

Frankly, meteorologists still do not know as much about jet streams as they would like, and predicting their existence, their location, and their strength is not an exact science.

The first evidence of high-velocity, high-altitude winds was noted in 1922 when an English weather balloon fell to the ground in Germany, 600 miles distance, only four hours later. The balloon had obviously made 150 miles an hour during its trip. In this instance, the meteorologists were mystified, for the surface winds over northern Europe were only normal—20 to 30 knots. So the episode was dismissed as a freak and isolated event.

In the 1930's, some theoretical investigations were made which indicated that very fast winds were occasionally found in the general west-to-east flow of air at high altitudes, but weathermen knew very little about their significance or frequency.

In 1944, however, the same freak winds were noticed again, this time by many B–29 pilots flying bombing missions between Saipan and Tinian northwestward across the Western Pacific to Japan.

On these occasions, pilots and navigators found the head winds at 35,000 ft so strong they were unable to reach Toyko and still have sufficient fuel to return to Saipan. Some of the B–29 navigators calculated winds in excess of 200 miles per hour! At first, the weathermen could not explain and did not believe such wind speeds were possible. But it soon became apparent that indeed they did exist.

This experience and knowledge spurred meteorologists to learn more about the structure and behavior of the wind patterns in the higher portions of the atmosphere. Aided by newer weather instruments for measuring wind conditions aloft and pilot reports from an increasing number of aircraft operating at high altitudes, meteorologists were able to establish a number of important facts about jet streams.

The Causes of Jet Streams

The exact causes of the jet stream are not yet completely understood. It is known they are closely associated with (1) the tropopause and its location, (2) the polar front and the associated outbreak of cold, polar air into the middle latitudes, and (3) low-pressure systems. Some visual indication of their presence can be seen in high, wispy cirrus clouds called "streamers."

To understand the causes of the jet stream, a pilot must understand the tropopause, since jet streams are usually found near it.

The tropopause, as we learned in Chapter 2, is the junction or boundary of the troposphere and the stratosphere. Over the North and South Poles, the tropopause is found at 25,000 to 30,000 ft. In the vicinity of the equator, however, the tropopause is found at 50,000 to 60,000 ft. At mid-latitudes—between 30° and 60°—there is sometimes a break in this boundary, as shown in Figure 19–1. It is in this boundary where pilots can encounter turbulence without clouds (clear-air turbulence), and the high-speed jet stream.

Thus, the jet stream can be described as a narrow, high-speed meandering river of air moving around the temperate zone in wavelike patterns, a part of the normal windflow aloft. Jet streams appear to grow, shift, intensify, decline, and finally die. They are sometimes continuous around the globe (see Figure 19–2) but more often they are broken up into several segments. The orientation, location, zones of maximum wind, and thickness of the jet core generally vary with latitude, longitude, and altitude.

Where Jet Streams Exist

In recent years, following voluminous observations of the upper atmosphere, it has been concluded that jet streams occur near regions where there is a large horizontal temperature difference between warm and cold air masses. That is, the jet stream velocity is greatest where north-south temperature contrasts are greatest. Although the large horizontal temperature difference is well below the tropopause, usually at the 20,000-ft level, the jet stream is closer to the tropopause. The region of temperature difference is usually associated with a cold front or cold outbreak aloft. The jet stream is located in the warm air

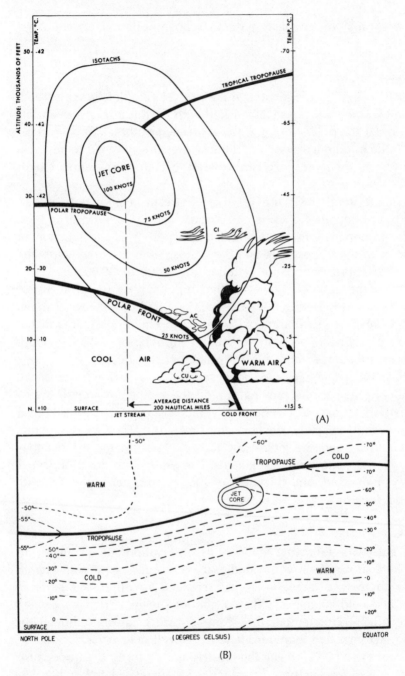

Figure 19–1 Vertical Cross Section of a Model of the Jet Stream. (A) Winds and fronts. (B) Temperature distribution.

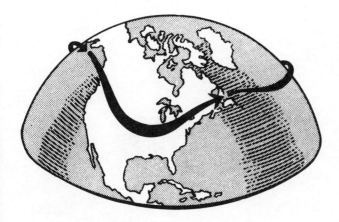

Figure 19–2   The Jet Stream Can Circle the Globe.

with the jet core of maximum winds located in the vicinity of 35,000 ft.

Wind Speed in Jet Streams

Wind speeds in a jet stream can reach as high as 300 knots. Two hundred fifty knots have been recorded (see Figure 19–3). Generally, however, the jet stream is composed of winds between 100 and 150 knots, and only in extreme cases, 200 to 250 knots. The high velocity of the jet stream extends for a great distance along the length of the stream and is composed of winds between 100 knots on either side of the central core of maximum winds. Wind speeds decrease very rapidly to the north (polar side), and more slowly to the south, or tropical side, of the stream. The greater the distance from the core of the jet stream, the lesser the wind speed. In fact, 100 miles north of the core, the wind may drop off as much as 100 knots, and as little as 25 knots in 100 miles of its southern edge. In terms of altitude, a decrease of 30 to 40 knots in 1000 ft above or below the core of the maximum winds is not uncommon.

Therefore, if a pilot finds that he is in the jet stream and fighting an adverse headwind, he should either climb or descend, *toward the cold air mass*. In this manner, the decrease in the headwind velocity will be much more abrupt than if he had altered his course to the *tropical* or warmer side of the stream.

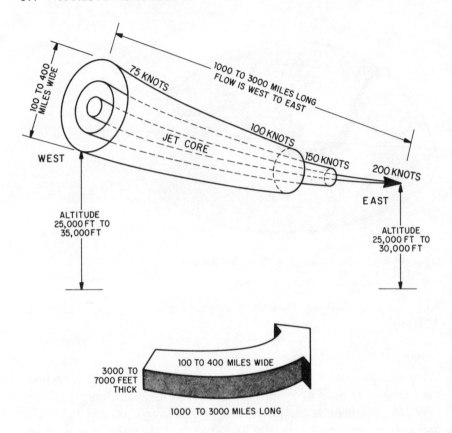

Figure 19-3 The Jet Stream.

General Characteristics of the Jet Stream

As previously mentioned, the jet stream can be a continuous river around the temperate zone, but more often it is broken into a series of well defined and disconnected segments. Moreover, there may be two or more jet streams in existence at the same time. For example, one may find a jet stream along the northern portion of the United States and another well-defined stream along the southern states (see Figures 19–4 and 19–5).

The frequency of jet streams does not seem to have any marked seasonal variation—generally as many occur in summer as in the winter. However, jet streams have a seasonal north-to-south shift and are more frequently encountered in the winter in the temperate zones than in summer. In the winter, in fact, jet streams are frequently

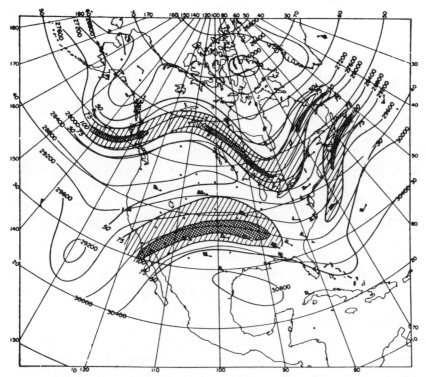

Figure 19–4 A Synoptic Example of the Wind Field at 300-mb on October 25, 1950, 0300 GCT, showing the Multiple Structure of the Jet Stream. Hatched areas denote regions with wind speeds exceeding 75 knots, cross-hatched areas have wind speeds exceeding 100 knots.

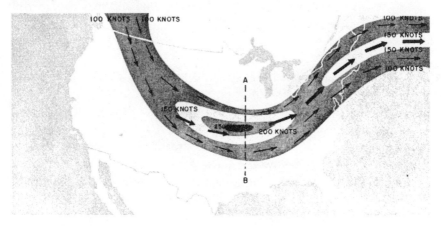

Figure 19–5 A Cross Section through a Jet Stream Showing the Core of Very Fast Winds.

located as far south as 20° north. The height of the jet stream is not uniform, and there is evidence that the core of maximum winds rises and falls in altitude as the stream moves erratically.

From day to day the jet stream develops in intensity and length, and tends to migrate, and its daily position is best determined from weather observations and pilot reports.

As the jet stream moves south, it appears that the speed in the core increases. In other words, if two jet streams exist over the United States, the more southerly one will generally contain stronger winds. Moreover, the wind speeds are stronger in winter than in summer.

The well-defined segments of the jet stream are normally 1000 to 3000 miles in length, 100 to 400 miles wide, and 3000 to 7000 ft thick (see Figures 19–3 and 19–5).

Aircraft flying with the jet stream will find their speed and range greatly increased, and conversely, those flying against the stream will find their speed and range greatly reduced.

The Subtropical Jet Stream

In winter there is frequently a secondary jet in a more southerly latitude than the polar front jet stream, near latitude 25° north to 30° north. Frequently this jet can be observed to extend from the Hawaiian Islands eastward to southern Florida with winds of 100 to 120 knots at 30,000 ft. This jet stream does not remain fixed from day to day, although it is remarkably persistent from time to time. Sometimes it seems to drift northward over the United States and may merge with the mid-latitude jet.

Jet Stream Summary

(1) The jet stream must be at least 50 knots and can be as much as 250 (verified) and 300 knots (estimated).
(2) Speeds of 200 knots over the United States in the winter are common; in the summer, they fall to 75 to 100 knots.
(3) The sky is usually clear at the jet stream core level, but there is often cloudiness below the jet stream. Cloud formations around the outer portions of a jet stream are visually wispy, streaked, and drawn-out in the direction the wind is blowing.
(4) A well-defined jet stream is normally 3000 to 7000 ft thick, 100 to 400 miles wide, and 1000 to 3000 miles in length. (See Figure 19–5.)

(5) Two jet streams can exist over the United States. If two are present, the southern one will usually have the strongest winds.

(6) There are as many jet streams in the summer as there are in the winter.

(7) Jet streams are thought to be important in predicting the weather. Because of their great speed and force, it is believed they help make weather conditions underneath them. By tracking the path of a jet stream, meteorologists can sometimes tell where it will go next and what kind of weather it will bring.

Flying the Jet Stream

From the above discussion, it is obvious that a pilot flying *west to east*, if his airplane can fly high enough, should endeavor to take advantage of the jet stream. The airlines are particularly sensitive to the advantages of the jet stream. In fact, one airline on the Tokyo to Honolulu route saved a half-million dollars in fuel costs, not to mention en route time, during a series of 220 flights by utilizing the jet stream eastward, and avoiding it westward.

Just as important and obvious, of course, is that the high-flying pilot going *west* should avoid the jet stream.

How does one do so? By consulting the upper-altitude wind charts that the meteorologists prepare and by reading pilot reports (PIREPs). Also, ground controllers will endeavor to assist pilots flying eastward in taking advantage of jet streams, and pilots flying west to avoid the jet streams.

CLEAR-AIR TURBULENCE

One of the major hazards to modern, high-performance aircraft is the problem of clear-air turbulence—a rough cobblestone type of bumpiness experienced in cloudless portions of the sky usually above 15,000 ft, with a maximum occurrence near 30,000 ft. This bumpiness, occurring without visual warning, is certainly uncomfortable and can possibly cause serious aircraft stresses. CAT areas are not accurately predictable in both place and time. Most instances of pronounced CAT are associated with the jet stream, or more specifically with abrupt increases or decreases of wind velocity with altitude, and are experienced more frequently during the winter months when jet-stream winds are strongest. The presence and intensity of CAT are not constant along the entire jet stream, however. Statistically, CAT usually occurs with wind gradients in excess of eight knots per thousand feet. Because of its random and transient nature, exact locations of clear-air turbulence are extremely difficult to forecast.

The two most probable locations near the jet stream for the occurrence of CAT are (1) just below and to the left (back of the wind) of the jet-stream core (between 22,000 and 28,000 ft), and (2) directly above the core in the vicinity of the tropopause (between 35,000 and 50,000 ft) for a considerable distance, perhaps 200 miles. The most abrupt wind shears* are generally found in these regions. CAT does not occur *in* the core, however. (See Figure 19–6.)

The occurrence of CAT can also extend to very high altitudes and can be associated with other wind-flow patterns. For example, a sharp trough aloft, especially one moving at a speed greater than 20 knots, can have clear-air turbulence in or near the trough, even though the wind speeds can be rather low as compared with the speeds near the jet stream. However, the winds on opposite sides of the trough can have a difference of 90° or more in direction (see Figure 19–7). CAT can occur in the circulation around a closed low aloft, particularly if the flow is merging or splitting, and to the northeast of a cut-off low aloft (see Figure 19–8).

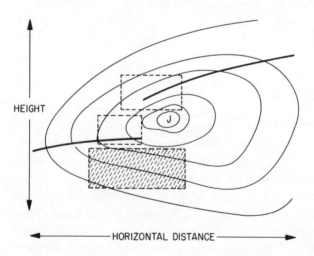

Figure 19–6 Areas of Probable Clear-air Turbulence in the Jet-stream Model.

* *Shear* is defined as the change in velocity (speed and direction) with distance measured in a direction perpendicular to the flow.

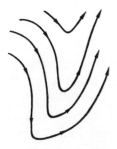

Figure 19-7   Clear-air Turbulence Associated with Sharp Trough.

When a pilot anticipates or encounters CAT, he should fly at a speed approximately 50% greater than stall speed for his particular altitude. Ordinarily, this will reduce turbulence which is of moderate intensity to light or only perceptible intensity. However, if the intensity of the turbulence is such that further action is required, the pilot should consider a climb or descent and/or turn to either side of the turbulent zone, using information provided by the weather forecaster during preflight briefing or over pilot-to-forecaster facilities. In such a situation, you should make very gradual climbs, descents, and turns.

Needless to say, pilots flying in or near the jet stream should familiarize themselves with the available weather reports and forecasts concerning CAT and avoid areas where it is reported as severe.

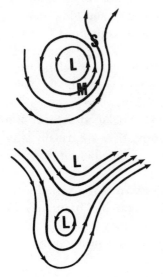

Figure 19-8   Upper: Clear-air Turbulence associated with Closed Low (M-merging flow; S-splitting flow). Lower: Clear-air Turbulence Associated with Cut-off Low.

CONTRAILS

A *contrail* is defined as a cloud-like streamer generated in the wake of aircraft flying in clear, cold, humid air. There are three types: (1) exhaust trails, (2) aerodynamic trails, and (3) dissipation trails.

Contrails are formed when aerodynamical cooling of the air or the addition of water vapor to the air creates a condition in which the air contains more than the maximum amount of water vapor it can hold at the existing temperature.

Aerodynamic trails are caused by the momentary reduction of pressure of air flowing at high speed past an airfoil in air that is clear and almost saturated. These trails are usually of short duration. They form at the tips of wings and propellers during times of extreme flight maneuvers such as sharp "pullouts" and high-speed diving turns. They occur at levels where the atmosphere is near saturation. If they do form, a pilot can make a small change in altitude or a reduction in airspeed and stop their formation.

Engine exhaust trails arise when the water vapor in the exhaust gas from an aircraft engine mixes with and saturates part of the air through which the aircraft is passing.

The fuel used in airplanes, both conventional and jet, is a hydrocarbon which, upon combustion, results in the addition of water vapor and heat to the wake of the aircraft. The formation of contrails of this type is a function of relative humidity, pressure, and temperature. For a visible contrail to form, the wake of the aircraft must first become saturated.

Studies of contrail formation indicate that there is a relationship to latitude, altitude, and season. The greatest general probability of contrail formation is in January, and the least in July.

When cirrus clouds are present, contrails usually form near the cirrus level. In this case, when the contrails form, they are usually very persistent, but may be difficult to distinguish from the surrounding clouds.

Investigation of the contrail problem has shown that a variation in engine power setting has no significant effect on the frequency of contrail formation. However, there is some indication that an increase in power setting probably results in an increase in trail intensity.

HAZE LAYERS

In the upper troposphere there frequently exist haze layers which are not visible to ground observers and do not appear to be in the nature of ordinary cirrus clouds. The bases of the haze layers are less well defined than the tops. Although the visibility above the haze is excellent, air-to-air and air-to-ground visibility in the haze layers is sometimes reduced to zero, depending on the position of the sun relative to the pilot. This high-level haze will not usually be present with a fresh polar outbreak, or if an air mass is on the move. Generally, high-level haze occurs with stagnant air masses.

CANOPY STATIC

Canopy static is similar to precipitation static encountered at lower levels. The solid particles which brush against the canopy or other particles which brush against the canopy or other plexiglass-covered surface of aircraft during flight, build up a static electric charge on them. When this static electricity is discharged onto a nearby surface or into the air, the accompanying noisy disturbance reduces radio reception. This charge and discharge of electricity can occur in rapid succession and appear as a more or less continuous disturbance. The ice crystals of cirrus clouds are the primary producers of canopy static at high altitudes.

If you encounter it, you can usually eliminate it by changing altitude.

20 SOARING AND METEOROLOGY*

INTRODUCTION

This chapter is an introduction to soaring meteorology, especially for beginning and intermediate pilots. In the broadest sense, the soaring pilot uses meteorological information in three ways: first, to enable him to forecast the expected conditions, perhaps a day in advance, so that he can plan his flying activity to take advantage of them; second, to understand the existing conditions which prevail at the start of the flight; and third, to enable him to recognize and properly interpret the dynamic changes in the meteorological situation which may occur during flight. (A list of references is given at the end of this chapter

* This Chapter is a condensation of the *American Soaring Handbook*, by Dr. Harner Selvidge, a respected and recognized expert in the soaring field. Dr. Selvidge is also a professional member of the American Meteorological Society. The author is grateful for his permission to include this condensation of his work.

for those who wish to delve more deeply into the science of soaring.)

THERMALS

A sailplane can stay aloft only when it is flying in a parcel of air which for some reason or other is rising faster than the sinking rate of the aircraft. The most important and widely used rising air currents in modern soaring are those caused by thermal effects, and are aptly called *thermals*. Their nature and characteristics are of vital importance to the soaring pilot. If we heat a parcel of air to a temperature higher than the air around it, it will rise because hot air is lighter and less dense than cold air. Just how far and how fast it will rise depends on a few simple meteorological factors which are discussed in this section.

The heat, which is the driving force behind the thermal, comes mainly from the sun. The soaring pilot should be on the lookout for special local heat sources such as brush fires, power plants, large factories, etc., but in general there will be no thermals if the sun does not reach the earth. The earth's surface is not uniform, and some spots get heated more than others. For example, a black area will absorb more heat than a light area and its temperature will be correspondingly higher. Air which is in contact with this warmer patch will be heated more than nearby air and will tend to rise. As it rises, it is replaced with cooler air from the surrounding areas which air is in turn heated, and the cycle is self-perpetuating. In its simplest form, this is the way a thermal is established over land. However, even if the surface is uniform, as is the case over large bodies of water, thermal cells form in fairly regular patterns, but much stronger thermals are found over land because of its surface irregularities.

Thermals—Characteristics of Rising Air

Let us now examine this heated parcel of air as it starts to rise from the earth. As soon as it leaves the vicinity of the surface its temperature no longer increases. The sun shines right through air without heating it appreciably. The only way it can pick up heat is by conduction from the hot ground. Furthermore, as it rises, the parcel of air expands as the pressure falls with increasing altitude. As a gas expands, it gets cooler. When dry air expands, it gets cooler. When dry air expands by being lifted to a higher altitude without exchanging heat with its

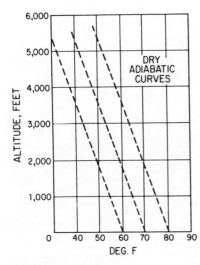

Figure 20-1

surroundings, it cools at a rate of 5.5°F. per 1000 ft (1°C. per 100 meters). The rate of cooling is called the *dry adiabatic lapse rate* by meteorologists. A family of curves showing this effect is given in Figure 20–1. For example, if a parcel of air with a temperature of 60° at ground level were raised 2000 ft, it would have a temperature of 49°. If it were 80° at the ground and the air were lifted 4000 ft, its temperature would be 58°.

Now let us see what happens in real life. Figure 20–2 shows a

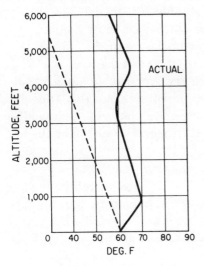

Figure 20-2

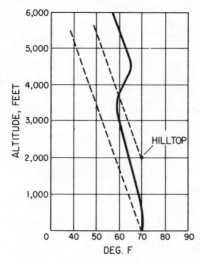

Figure 20–3

similar diagram, but plotted on it in a solid line is the actual curve of temperature of the air versus altitude at that time of day. The time is in the morning, and the ground, having lost heat by radiation at night, is at 60°. Since in "normal" conditions the air gets cooler at higher altitudes, a situation where the air gets *warmer* with altitude is called an *inversion* condition. In this case we have an inversion at ground level with a top at about 800 ft, and another inversion starting at 3500 ft with a top at about 4500 ft.

Suppose we took a parcel of air at the ground temperature of 60° and mechanically raised it to higher altitudes. Its temperature would follow the adiabatic law shown by the dotted line. Note that at every altitude it would be *cooler* than the temperature of the surrounding air (given by the solid line). For example, at 2000 ft the temperature of our raised parcel would be 49° while the air surrounding it is 64°. If air is cooler than the surrounding air, it will sink, not rise. Thus this morning as shown we have a very stable condition, and there will be no thermal activity, since air at ground temperature or slightly higher cannot rise of its own accord.

Figure 20–3 shows the situation a couple of hours later when the ground temperature has risen to 70°. The low-level inversion has almost been wiped out by the heating of the ground by the sun, and the subsequent transfer of this heat to the lower layers of air. A parcel of air at ground temperature of 70° still cannot get up, but let's look

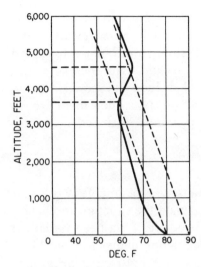

Figure 20–4

at the top of a 2000-ft hill nearby. The temperature of the surface of the ground at the hilltop will be about 70°, the same as that in the valley, since it gets about the same radiation from the sun. If we watch this parcel of 70° air originating at the hilltop, it will rise along the adiabatic line shown until it intersects the air-mass temperature curve at about 3700 ft. Thus there will be thermal activity on the hilltop before there is any in the valley. Whenever the line representing the adiabatic lapse rate slopes to the right of the air-mass temperature curve, we have convection and an unstable condition at those altitudes.

When the temperature has reached 80° on the ground as shown in Figure 20–4 the soaring will be good, with the top of the lift at about 3700 ft. If the temperature at the ground should reach 90°, lift will be good up to about 4700 ft. This sequence of events shows how the forecaster predicts thermal conditions. It requires radiosonde data on the air-mass temperature at all altitudes, which will be available only at the Weather Bureau office.

The maximum thermal strength is usually found in early or mid-afternoon, when the sun has had time to thoroughly warm up the ground. Later in the afternoon the sun strikes the ground at a considerable angle and with a much longer path through the atmosphere, so the heating is less and the thermals start to weaken, usually vanishing before sunset. As shown above, the tops of thermals may get higher as the day wears on, and it is usually observed that they increase in strength and become further apart. Surprisingly, air with water vapor in it is lighter than dry air. Thus thermal strength can be partly caused by water vapor in the air, as well as by heating. If the thermal carries the water vapor to the condensation level, a cloud will form. However, there can be very strong "dry thermals" on days when there are no clouds at all. Lift and its relation to clouds is discussed in a later section.

Structure of Thermals

There has been considerable theoretical and experimental work performed in an effort to understand and explain the structure and action of thermals. Various possible thermal structures have been described by meteorologists, aerodynamicists, and soaring pilots, and it is likely that more than one kind exists in nature. The most common experience of soaring pilots is that, if they fly in under another climbing sailplane, they catch the same thermal, even though they may be as much as a couple of thousand feet below the other aircraft. This seems to indicate that in a vast majority of cases an assumption of a thermal structure like a column or chimney best fits the actual case. The dust devil is a common visual manifestation of this phenomenon.

A favored theoretical thermal structure is the so-called vortex shell, which is like a smoke ring blown straight up as in Figure 20–5. The air in the middle is rising faster than the ring as a whole. A sailplane could be supported and climb in the middle of the vortex ring, but one flying below will find no lift. While most soaring pilots have occasionally experienced a situation where the thermal seemed to be composed of a discontinuous series of bubbles of lift which might be vortex shells, this occurs in only a very small percentage of the time

Figure 20–5 Vortex Shell.

in the opinion of most soaring pilots. However, a long drawn-out vortex shell or bubble would resemble a column, so there is probably a very indistinct dividing line between the column and vortex ring concepts.

If the wind velocity increases with altitude, the thermal column will be deflected. Its top will incline downwind, whether the thermal roots are tied to a single spot on the ground, or whether the column is drifting with the wind. In these circumstances a pilot coming in under a sailplane higher in a thermal will generally find the lift upwind of the other ship as shown in Figure 20–6. On any particular day, the thermals will have surprisingly uniform characteristics in strength, cross section, and frequency. The strongest lift will generally be found in the center of the thermal, with this area of good lift growing larger as you get higher in the thermal, so circle tightly to get the best lift. One additional point about thermals should be mentioned. With all this air going up, it must come down somewhere. It does this between the thermals. However, the areas of down are many times larger than the areas of lift, so the downward velocity outside the thermal is less than the upward velocity in the thermal center. This effect may sometimes be masked if the air mass over a several-square-mile area is upwelling or subsiding, but it is always well to remember that the stronger the lift in the thermals, the stronger will be the down currents you may encounter between them.

Do thermals rotate? This point has by no means been settled, but it probably is of little practical importance to the soaring pilot. Lift will be better if he circles against the direction of rotation, but without a point of reference the pilot cannot be expected to tell the

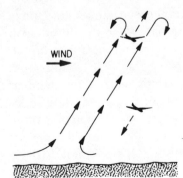

WIND

Figure 20–6 Thermal Column.

direction of rotation. Circle in the direction which is most comfortable to you, unless the pilots who have preceded you in the thermal have established a different direction. In the case of dust devils, there is a definite rotation which can easily be observed. There appears to be no preferred direction of rotation, both clockwise and counterclockwise being noted in about equal numbers. Dust devils may vary from a small swirl of wind along the ground, to a column of dust several thousand feet high that lasts for ten to fifteen minutes. The visible diameter of the dust devil is usually too small for the soaring pilot to circle in, but the lift around it will be very strong and frequently violently turbulent. Don't try to fly in a dust devil below 500 ft, or you may not have enough altitude to recover if you get thrown out of the core in a full stall. No more is known about dust devils than thermals, but at least you can see them.

Where to Find Thermals

The thing that the soaring pilot has foremost in his mind about thermals is, "Where is the next one?" As pointed out earlier in this section, thermal sources are generally hot spots on the surface of the ground. What places get hotter than others? This is not an easy question to answer. If it were, everyone would have diamond badges, and no one would land unexpectedly. One fundamental point is that dry areas get hotter than moist ones. When there is much moisture present, so much energy from the sun is expended in evaporating the moisture that the surface doesn't get very hot. Dry fields or dry ground of any nature are better thermal sources than moist areas. This applies to woods or forests, which are poor sources of thermals because of the large amounts of moisture given off by the leaves.

The more heat that reaches the surface of the ground, the greater will be the temperature rise. Thus when the sun strikes the sunny side of a hill at a greater angle than in the flat valley, the hillside will be a better thermal source. For a given ground temperature, the air temperature will increase more, the longer the air is in contact with the ground at that point. Strong surface winds mean poor thermals since the wind moves a larger volume of air across the hot spots, resulting in less heat transfer per unit volume of air. On the other hand, any heated area that is protected from the wind, like the lee slope of a hill, will have a better chance of producing thermals. These are sometimes called *wind shadow thermals*. As previously pointed out in Figure 20–3, air originating at the tops of hills or mountains is frequently in a region of greater instability than exists at lower levels, and the frequently quoted advice, "Keep to the high ground," is well founded.

It is often possible to find thermals on the windward sides of hills and ridges. When slope soaring in the updrafts near a ridge or hillside, constantly be on the alert for thermals which may be forming on the hillside below you. They will usually be found just upwind of the best ridge lift. If the increased lift is encountered at some point along the slope, try a circle into the wind to see if a thermal has started there.

It should be borne in mind that there are two conditions required for thermals: first, an unstable air mass above the ground; and second, some parcels of air near the ground which are relatively warmer than the surrounding air. Note that it is not required that the air anywhere be hot. Thus it is possible to have good thermals in the dead of winter. The author has seen thermal soaring flights of over an hour with the ground covered with six inches of snow. These thermals were possibly triggered by buildings, roads, or other temperature discontinuities.

When near the ground, one should always be on the alert for visual clues to the presence of thermals. A swirl of wind indicated in a growing crop, or in the tops of trees, is likely the start of a thermal. Smoke rising, columns of dust or haze (sometimes better seen with colored or polarized glasses), and birds soaring are all good tips on what the air is doing. One of the best indicators is smoke or dust converging from different directions.

CLOUDS

There are some soaring pilots who say they don't mind flying on days when there are dry thermals and no clouds, because then they don't go off chasing every cloud in the sky only to find no lift under many of them. This highlights the difficulty of using our knowledge of the relation of clouds to thermals. Probably 80% of the thermal soaring in the United States is performed under conditions when there is enough moisture in the air to form clouds, so the ability to read clouds as thermal indicators is a most important skill. Every cloud is formed of either water droplets or ice crystals, or both. When moist air is raised until its temperature is low enough to condense the water in it, a cloud forms. Further cooling at higher altitudes can change the water drops to ice crystals.

Cumulus Clouds

These are the joy of the soaring pilot. They mark the tops of thermals, or where the tops of thermals *used* to be. When the thermal lifts warm moist air to cooler levels, a cloud of visible water drops will form when the temperature falls enough to condense the water in the air. This condensation level will mark the base of the cloud. The lift will continue up in the cloud, however, as long as the thermal keeps supplying moisture and energy to it. Thus, we see the growing cumulus bulging and puffing with its top getting higher and higher. Lift in the cloud will be increased by heat released when the moisture condenses to form the cloud droplets, and at the higher levels in the cloud when the water turns to ice, releasing the heat of fusion. Eventually, the thermal which is feeding the cloud will pause or die out. The cloud will still be there, but the unfortunate pilot arriving under it at about this time will find no lift, or sometimes downdrafts. With the removal of the energy source, the cumulus cloud will start to evaporate, which increases cooling and accelerates the process of evaporation. The cloud will soon fade into mere wisps and disappear.

Meteorologists have studied summertime cumulus clouds, and find that a good figure for the average life of a cumulus cloud is only 20 minutes. Some of these clouds will last only a few minutes, while others will last an hour or more. This points out the importance of picking the right cloud when looking for thermals. The important thing is to pick the *growing* cloud. Here it is a great help if the pilot will plan ahead. While circling in one thermal, he should be looking in the direction he wants to go in order to become familiar with the cloud patterns developing there. Some pilots have developed the knack or instinct that permits them to tell at a glance whether a cloud is growing or dissipating. After the first five minutes or so, a growing cloud has a firm, solid appearance all around its edges as new condensed water vapor pushes out the top and sides. The shrinking cumulus has a more ragged look around the edges as the drops evaporate and are not replaced. During the first few minutes of growth there is little difference in appearance between the new cumulus and the last wisps of a dying one. This emphasizes the importance to the pilot of having a picture in his mind of all the clouds in the direction of flight. If the wisps appear where there had been clear sky, head for them. If there was a larger cloud there a few minutes before, avoid that area.

Another way to tell whether a cloud is growing is to look closely at some interesting spot on the edge of the cloud, perhaps a wispy section. Then look at this same spot a minute or so later (for example, the next time around in the thermal) and see if it has become more solid, or is getting more ragged. This takes lots of concentration and is hard work, but that is the sort of thing that gives one pilot the edge over another. All too much of your soaring life will be spent on the ground. You can make good use of this time by watching the cumulus clouds, and trying to pick the ones that are growing or dissipating. This is excellent training, and you don't fall out of the sky if you miss your guess, but can try again.

When there are many good cumulus clouds in the sky, the distance from the base to the top of a cloud is an indirect measure of the relative strength of the thermal which is driving it, although primarily it indicates the strength of the lift *in* the cloud. Watching the vertical development of clouds will give you additional information on the trend of thermal strengths on cross-country flights. If the cloud tops are getting higher, expect stronger thermals; but don't necessarily go charging off to get under the cloud with the highest top. It may only be marking where the strongest thermal *used* to be, ten or fifteen minutes ago. If, however, you have a choice of going to an area where there are cumulus clouds (even though they appear to be dying) or to a clear area, pick the cloudy one. The cloudy area may be connected with some terrain feature, and the clouds and thermals may soon re-appear in the same place. If you can hang on nearby, do so until things start picking up ahead.

On some days when there is very little moisture in the air, clouds will be scarce or nonexistent. Under these conditions watch closely for any small wispy cumulus beginnings which might appear. These are some of the most reliable indicators of thermals.

Cumulonimbus

These are thunderstorm or rain clouds which represent a late stage in the growth of a long-lived cumulus cloud. If the convection in the cloud cells is large enough, the moisture is carried high enough to cause the formation of large drops which fall out as rain. The tops of these clouds will consist mostly of ice crystals, and when the stronger winds aloft blow these tops away from the main body of the cloud, the well-known "anvil top" is seen. This marks a well-developed cumulonimbus, one to stay out of if possible. The ice particles blown off the top sometimes form a sheet sufficiently large to cut off the sun's heat from reaching the ground near the storm, further decreasing the convection. During its active phases, the large, well-developed cumulonimbus is one of nature's most terrifying phenomena, particularly when seen close at hand. Watch out for strong downdrafts, hail, turbulence, and shifting winds in their vicinity.

Cloud Flying

This is not the place for a treatise on how to fly in clouds, but some comments might be made of what you may expect to find in the cloud. One of the most important points is the choice of cloud. *Don't* pick big black cumulonimbus with thunder and lightning. *Do* pick small white cumulus with only two or three thousand feet of vertical development. Remember, the cloud will be continuing its development while you are in it, so if it is growing, things will always be getting rougher so long as you are in it. Also bear in mind that, when you are cruising at about cloud base, you have a very poor view of the amount of vertical development of nearby clouds or the one above you.

Cloud Base

The sharply defined flat cloud base that we usually consider to be characteristic of a cumulus cloud is a much more fuzzy situation when seen close up. Frequently it is not flat, but dished up in the middle, with the base of the cloud in its center a couple of hundred or more feet higher than at the edges. Since the bottom is actually rather wispy, you will find you get into the cloud gradually. Lift is frequently very strong here, so don't count on being able to just go in a few feet "to try it out." It might be a couple of thousand feet before you can get out. You can literally be "sucked up into the cloud" despite your wishes if you play too close to the base of a good cumulus. Don't do it unless you are ready to go in. When a big thunderstorm has its base only a couple of thousand feet above ground level, it is easy to get large areas of almost zero visibility in heavy rain. If you have to come down from the cloud in this situation in turbulence and extensive downdrafts, there is a very good chance of hitting the ground before you see it. This is another reason for not tackling big thunderstorms.

Lightning in Clouds

The well-developed cumulonimbus is usually highly electrified, and you can expect to find lightning in it. Frequently when flying nearby you will hear the thunder before the lightning is visible. That is the time to pick another cloud. Lightning strokes can be from point to point in the cloud, from cloud to cloud, and from cloud to ground. Keep away from the areas where these strokes are seen. Many aircraft have been struck by lightning with only minor damage, but some, including sailplanes made of wood and metal, have been severely damaged.

Icing in Clouds

Usually the water drops in a cumulus cloud do not freeze at zero centigrade but remain in liquid form, sometimes until $-20°$ or $-30°$ are reached. However, if you fly through them when the temperature is below freezing, they will turn to ice when they hit the aircraft. Layers of ice will build up on the wings, ruining the airfoil shape, and will cover the canopy so you can't see out. The added load of the ice also increases the stalling speed and makes the aircraft harder to control. One of the greatest hazards is the possible freezing up of the controls or dive brakes, so they cannot be operated. The only thing you can do is to get out of the cloud and down to temperatures where the ice will melt off. Don't forget to start out *before* the ice gets too bad. It will be continuing to build up while you are trying to get out.

Hail in Clouds

In the opinion of many, this is the greatest hazard in cloud flying. The worst case is striking hail large enough to break the canopy. It will be practically impossible to control the aircraft with wind, hail, and rain in your face. Almost as serious is the damage done to the surfaces. Large hail can rip fabric to shreds, and beat metal skins into unrecognizable nonaerodynamic shapes. All these have happened to sailplanes. Imagine striking thousands of golf balls at 60 miles an hour. Hail that size is very common. Pea-sized hail sounds fearsome enough. When you hit hail, head for home as quickly as you can. It will usually be found near the top third of the cloud, above the freezing layer, but if it is large enough, or the down currents strong, it can fall out the bottom of the cloud and be encountered anywhere. You can best avoid hail by staying out of the big, black, tall thunderstorms, or out of clouds that might develop to these dimensions while you are in them. Hail also can fall outside a cumulonimbus; therefore, a pilot must be careful in flying near such clouds.

Turbulence

Every pilot who does much cloud flying sooner or later encounters severe turbulence. Yet with proper caution, it is possible to fly many hours in clouds without difficulties from turbulence. The power pilot finds much more turbulence than the soaring pilot, because he flies *through* the cloud, first striking the downdrafts at the edges and then the one or more up currents on the inside, and then the down currents on the other side. In many cases, if the soaring pilot gets well centered in the main lift area in the middle of the cloud, he can be going up at several thousand feet a minute without the slightest bit of turbulence. The greatest turbulence will be encountered near the top and the edges of the cloud. Flying a straight heading to the outside of the cloud is the best way to get out of turbulence, but remember, it will probably get worse as you near the edge. However, you must fly through it to get out. Like all other cloud-flying problems, turbulence is minimized by staying away from the big black ones. Sometimes the nice simple, single-celled cumulus you start in will merge with some other ones nearby while you are in it, so you should not be surprised to encounter several cells of up and down currents in a single cloud mass.

MECHANICAL EFFECTS OF WINDS

Probably every soaring pilot has, at some time or other, made use of the lift resulting from the vertical motion of the wind when it strikes an obstruction such as a mountain or ridge as in Figure 20–7. The extent of the area of lift in front of the ridge will depend on many factors, such as the shape of the slope and the strength and the direction of the wind relative to the slope. If the slope is too steep, such as a vertical cliff, there will be strong eddies near the face, with poor lift (Figure 20–8). If the slope is too flat there will also be little lift.

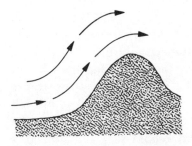

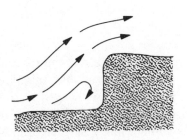

Figure 20–7 Gradual Slope.   **Figure 20–8** Sharp Slope.

The strongest lift will be found in front of a slope at right angles to the wind. As the wind strength falls the area of lift becomes narrower, and this is sometimes observed in flying, before an actual loss of altitude is experienced. Watch for a change in the location of the maximum lift as an indication of changes in the wind strength or direction.

Strong eddies and turbulence are frequently encountered on the lee or downwind side of hills and ridges. The greatest care must be exercised in flying or landing in such areas. Strong winds will give a standing wave in the lee of obstructions. Winds in the mountains are greatly influenced by mountain valleys or passes. Sometimes a venturi effect can multiply the force of the wind several fold. This is an added hazard to mountain flying in strong-wind conditions.

Up-Slope Winds and Drainage

In addition to the mechanical effects described above, there are important heating and cooling effects which have a strong influence on winds in hills and mountains. These effects are most noticeable where there is a valley with high ground on two sides. There is an early morning up-slope wind, caused by heating, which is of particular importance to pilots wanting to get an early start. A parcel of air close to the ground, halfway up the slope, will be heated more than a parcel of air at the same altitude out over the valley, and thus will rise faster. This causes an up-slope wind frequently capable of keeping a sailplane aloft before the heating is enough to set off regular thermals. This layer of lift is usually rather shallow, and it is frequently necessary to fly within one or two wingspans of the side of the mountain to take advantage of it, but it is very useful to experienced and alert pilots.

In the late afternoon and evening we have the drainage wind, which is the reverse of the up-slope wind. As the temperature of the higher-altitude ground falls, the air in contact with it is cooled more rapidly than air at the same altitude over the valley, and the cool air from the top of the hills flows down the slope to the valley. This is no time to be flying near the slope. Get out over the valley. In such circumstances it is not at all uncommon to have the air over the whole valley rising vigorously enough to keep a sailplane aloft long after sunset as the valley air is displaced by the cool air draining down from the high ground. These areas of lift are sometimes miles in extent. Watch out that you don't get caught up high and have to fight your way down to a night landing.

Wind Shears and Gradients

Friction with the ground slows the wind at lower levels, and one will almost always find that winds aloft are stronger than those on the ground. They sometimes blow in different, and even opposite, directions and there can sometimes be very significant changes in speed and direction within a hundred-foot change in altitude. This creates a dangerous situation when it occurs near the ground, since the sailplane can experience a drastic loss in airspeed at the moment it flies from a headwind to a tailwind. This is an important reason for carrying extra airspeed in strong-wind conditions near the ground.

Turbulence

Strong winds will not only create turbulence in thermal lift areas, but can also be so strong as to tear the thermals apart, making them very difficult to locate and to stay in when found. This problem will arise when the wind reaches 15 to 20 knots, and is primarily a surface phenomenon. The mechanical mixing of the air as the wind strikes objects on the ground will prevent the thermals from being well defined, and can make landing considerably more difficult. If you are flying in a strong-wind condition, you can be sure it will be turbulent near the ground, even if it is smooth aloft.

Wind Shifts

Sudden wind shifts in the vicinity of a storm can create hazardous conditions, if not noticed. In a typical situation, a pilot may be making a hurried landing in the immediate vicinity of a thunderstorm. As the storm matures and rainfall starts, enormous quantities of cool air pour down out of it and spread along the ground ahead of it. The direction of flow of this wind is not necessarily related to the general wind prevailing nearby. A very common sight under these circumstances is to see a runway with windsocks at each end pointing in opposite directions. Blowing dust will sometimes give a clue as to the wind direction. Watch out for where it appears to rise *up* along a line. This will mark the boundary between the two air flows at opposite directions.

Visible Thermal Eddies

A sudden change in wind direction at the ground on a relatively calm day usually means a thermal has moved by. This is frequently evidenced by eddies of dust and dry grass. If your aircraft is on the ground, watch out! All too many sailplanes have been overturned by such eddies while standing unattended. If you are in the air, and in that unhappy position of being only a few hundred feet from having to land, keep a sharp watch for swirls or eddies on the ground which will mark the birth of a thermal. These patterns can be noticed easily in dusty ground, but they are also obvious to the trained eye on bodies of water, fields of grain or high grass, and in the foliage at the tops of trees.

THE STABILITY INDEX

The Weather Bureau frequently uses what is called a Stability Index to describe an air mass, and although it is primarily used by the Bureau for the quick evaluation of thunderstorm potential from individual upper air soundings, soaring pilots will find it useful. It is computed as follows.

Using temperature and dew point at 850 mb (about 5000 ft MSL) the air parcel is "lifted" on the adiabatic chart up the dry adiabatic line to the saturation level and then up the saturated adiabat to the 500-mb level (about 18,400 ft MSL). The resulting temperature at 500 mb obtained from this process is then compared with (subtracted from) the actual temperature measured by the upper air sounding. If this hypothetical lifting process, which could actually be triggered by surface heating, produces a temperature at 500 mb which is warmer than the actual sounding, the air is unstable and the lifted parcel of air would continue to rise. Such a condition would produce a negative

index value and would be a factor which favors thunderstorm development. If the same lifting process should result in a temperature cooler
than the sounding at 500 mb the index would be positive, the "lifted"
air would then sink, and the air would be unfavorable for thunderstorms.

The Stability Index is computed and reported by each upper air
station and is available to most pilot weather briefing stations. If you
are contemplating a cross-country flight, be sure and ask about the
stability in the direction you are heading. Unfortunately, since the
Stability Index is designed for thunderstorm, not thermal, prediction,
it can sometimes be deceptive. Its scale runs from about minus seven,
for heavy thunderstorm activity, to plus fifteen, for an extremely
stable condition. Negative numbers will almost always mean good
soaring, but if the air mass is very dry, the Stability Index might be
plus five or six, indicating no chance of thunderstorms, but there
might be good soaring in dry thermals.

A variation of this index has been devised for soaring use, eliminat-
ing the moisture factor. It is a scale of five as shown below:

> Plus two —Stable; no thermals
> Plus one —Weak thermals
> Zero   —Dry and unstable; good thermals
> Minus one—Cumulus clouds
> Minus two—Thunderstorms

This scale has been used for soaring contests, but it is *not* the one you
will get from the Weather Bureau.

While on the subject of stability, you should ask the forecaster
about the altitude of cloud base, if any are forecast, and possible
height of cloud tops. The vertical development of a cumulus cloud is
a good clue of thermal strength to be expected.

WIND FORECASTS

All weather stations have wind forecasts for standard levels of 5, 7, 10,
14, 18, and 24 thousand feet ASL. The strength and direction of the
winds aloft are usually the major factors in determining the direction
of flight for cross-country soaring. A strong tail wind is nice, but remember, if the surface winds get up to 25 to 30 knots they may start
tearing the thermals apart, and make thermal soaring very difficult.

Of course, for ridge and wave soaring you want strong winds, but in this case too, you want to know the velocity and direction.

TEMPERATURE

Temperature at various altitudes is forecast like the winds, and in some Weather Bureau stations it is measured by balloon-borne radiosondes. One of the questions you will wish to ask the forecaster is the ground temperature at which thermals will first be triggered. The forecaster can read this off his reports of radiosonde soundings. Be sure to tell him the altitude of the field if it is different from that of his station. You would like to have him tell you at what time of day this temperature will be reached, and he might be willing to hazard a guess, but this is greatly influenced by local effects so don't expect it to be very accurate. While you are talking temperature, ask the maximum predicted for the day. When compared with the triggering temperature this will give some idea of how long the thermals might last in the afternoon.

TRENDS

It cannot be emphasized too much that one of the most useful tools in short-range forecasting is to have a good idea of the expected trend in the weather. For example, a forecast of a front coming through at 9 P.M. might seem of no interest to a soaring pilot who is sure to be down by 6 P.M. But if the wind suddenly changes direction about four or five o'clock, it is probably a good sign that the front has speeded up and is arriving ahead of schedule. You must always be observing the immediate conditions and must update your forecasts accordingly. Once you cut your communications with the forecaster, you are on your own, but an amateur can do surprisingly well in a changing situation if he knows in advance what might be expected. Forecasts of the *sequence* of expected meteorological events are usually very good. Forecasting the exact *timing* is the difficult part.

NOTE: This chapter is only a condensation of Dr. Selvidge's *American Soaring Handbook* published by the Soaring Society of America, Incorporated. Cost is $1.00. For those soaring pilots more interested in soaring meteorology, the book *Meteorology for Glider Pilots*, by C. E. Wallington and published by Murray, London, is available through Schweizer Aircraft Company, Elmira, New York. Cost is $8.00.

APPENDIXES

APPENDIX 1

GLOSSARY

ABSOLUTE CENTIGRADE—A scale of temperature whereon 373° represents the boiling point of pure water at standard atmospheric pressure, and 273° represents the melting point of pure-water ice. It differs from the centigrade scale only by the constant 273°. For conversion of centigrade to absolute centigrade, see page 29. (See also STANDARD ATMOSPHERE.)

ABSOLUTE ZERO—The temperature of a body from which all heat has been removed, and at which all molecular action ceases. For practical purposes, it is −273°C. and −460°F. It is physically impossible to have any degree of temperature below "absolute zero," which, though impossible of practical attainment, marks a very real and practically useful basis for calculations involving the laws of gases.

ACTIVE FRONT—A front which produces considerable cloudiness and precipitation.

ADIABAT—A curve or line on a chart showing the adiabatic lapse rate, whether dry or moist, respectively.

ADIABATIC PROCESS—A process in which a mass of air (or other gas) is compressed or expanded in volume without gaining or losing any heat to or from outside sources. Its temperature and volume will change in a manner peculiar to such circumstances only. In expansion it will undergo *adiabatic cooling,* while in compression it will undergo *adiabatic heating,* and its change in temperature will be an *adiabatic temperature change.* When air is lifted it expands and cools adiabatically at a rate of 1°C. for each 100 meters (or 1°F. for each 180 ft) of altitude change, if the air is unsaturated; this is called the *dry-adiabatic lapse rate* of temperature change. If the air is saturated, the air cools at only (about) one-half the above rate, and this is known as the *moist-adiabatic lapse rate*. The reason for the difference between these two lapse rates is that, in the condensation of saturated air, the *latent heat of vaporization* is released back into the air, tending to reheat the air to a certain extent, thus slowing down the rate of cooling of the saturated air.

ADVECTION—Process of transfer by means of horizontal motion.

ADVECTION FOG—Fog formed when warm moist air comes in contact with a colder surface. This contact with the cold surface cools the air to its dew point, thereby causing condensation and fog. It is a common phenomenon throughout the year over cold ocean currents (for example, the Labrador current) when the air comes from warmer regions.

AEROLOGY—That branch of meteorology which treats of the free atmosphere, that is, unaffected by surface effects.

AIR—A mechanical mixture of gases surrounding the earth and of which the earth's atmosphere is composed.

AIRFALL—A current of heavy air flowing down over the tops of ridges, buildings, etc.

AIR MASS—A term applied by meteorologists to, an extensive body of air within which the conditions of temperature and moisture in a horizontal plane are essentially uniform.

AIR POCKETS—Areas in which descending currents of air cause an aircraft to drop rapidly toward the earth. They are found principally in clouds of the cumulus type, over small water or wooded areas, and on the lee side of hills, mountains, hangars, and other obstacles.

ALTIMETER—An instrument based on the principle of the aneroid barometer and used for indicating altitude above or below a given datum point, usually the ground or sea level.

ALTIMETER SETTING—Pressure of the reporting station converted to make the altimeter read zero elevation at an altitude of 10 ft above mean sea level (or to read field elevation 10 ft above the runway).

ALTIMETRY—In Pressure Pattern Flying, the use of comparative readings of two altimeters of different types (that is, one of the radio absolute type, the other of barometric pressure type) to obtain an indication of horizontal pressure gradient while in flight.

ALTITUDE—Vertical distance measured above sea level. (See also PRESSURE ALTITUDE.)

ANEMOGRAM—The record traced by a self-registering anemometer.

ANEMOGRAPH—An instrument which makes a running record of the velocity of the wind.

ANEMOMETER—An instrument which measures and indicates the momentary velocity of the wind.

ANEMOSCOPE—An indicating instrument with an attached wind vane that shows the wind direction on a calibrated scale.

ANEROID BAROMETER—An instrument for measuring the pressure of the atmosphere. Variation of the air pressure causes the expansion or contraction of a sealed thin, flat, metallic cylinder which in turn causes a pointer to move across a graduated dial. Temperature changes and mechanical friction within the instrument require that it be checked frequently with a mercury barometer and corrections made accordingly.

ANTICYCLOGENESIS—The term applied to the process which creates or develops a new anticyclone. The word is applied also to the process which produces an intensification of a preexisting anticyclone.

ANTICYCLONE—An area of high pressure.

ANTITRADES—Upper tropical winds blowing steadily in a direction opposite to the trade wind, beyond which, in the North Temperate and the South Temperate Zones, it becomes a surface wind.

ANVIL—The characteristic fibrous, spreading top of a cumulonimbus cloud in full development.

AQUEOUS VAPOR—Water vapor.

ARCTIC AIR—Air of an air mass whose source lies in the frigid arctic or antarctic regions.

ARCTIC FRONT—The line of discontinuity between very cold air flowing directly from the arctic regions and polar maritime air that has moved away from its source region in a more or less circuitous path and been warmed through contact with the ocean surface.

ARCTIC SMOKE—A form of fog caused by rapid evaporation from the surface of warmer water when very cold, dry air streams across it.

ATMOSPHERE—The whole mass of air surrounding the earth.

ATMOSPHERIC PRESSURE—The force per unit area exerted by the weight of the atmosphere from the level of measurement to its outer limits.

AURORA—A luminous phenomenon due to electrical discharges in the atmosphere; probably confined to the tenuous air of high altitudes. It is most commonly seen in subarctic and subantarctic latitudes. Called *aurora borealis* or *aurora australis,* according to the hemisphere in which it occurs. Observations with the spectroscope seem to indicate that a faint "permanent aurora" is a normal feature of the sky in all parts of the world.

BACKING—A shift of the wind in a counterclockwise direction, that is, to the left of the direction from which it had been blowing. (See also VEERING.)

BAR—A unit of pressure equal to one million dynes per square centimeter; 1 bar = 100 centibars = 1000 millibars. A barometric pressure of one bar is sometimes called a *C.G.S. atmosphere* and is equivalent to a pressure of 29.531 in. of mercury at 32°F. and in latitude 45°.

BAROGRAM—The continuous record made by a self-registering barometer.

BAROGRAPH—An instrument based on the principle of an aneroid barometer which makes a continuous record of atmospheric pressure.

BAROMETER—An instrument which measures atmospheric pressure.

BAROMETRIC TENDENCY—The changes of barometric pressure within a specified time (usually three hours) before the observation, usually indicated by symbols at the right of the station as shown on the weather map.

BEAUFORT SCALE—The scale of wind force devised by Admiral Sir Francis Beaufort in 1805, beginning with dead calm indicated by 0° and ending with hurricane indicated by 17. The scale is based on the effect of wind on various ground objects such as smoke, leaves, trees, etc. (Numbers above 12 are an extension of the original scale.)

BLIZZARD—A violent, intensely cold wind, laden with snow.

BUMPINESS—A condition of the air caused by thermal or mechanical turbulence, that is, ascending and descending currents and eddies, which produces a sensation in flight comparable to riding along a rough road in an automobile. Such a condition close to the ground may be dangerous and requires great caution and alertness of the pilot.

BUYS-BALLOTT'S LAW—In the Northern Hemisphere, if you face the wind, the area of lower pressure is on your *right* and (generally) slightly behind you. In the Southern Hemisphere, if you face the wind, the area of lower pressure is on your *left* and (generally) slightly behind you.

CALIBRATION—The name ordinarily given to the process of ascertaining the corrections to be applied to the indicated readings of an instrument in order to obtain true values.

CALM—A region where air movement is less than one mile per hour. Beaufort number 0.

CALMS OF CANCER; CALMS OF CAPRICORN—The belts of high pressure lying north of the northeast trade winds and south of the southeast trade winds, respectively.

CASCADE—The name applied to the mass of spray or dense vapor thrown outward from around the base of a water spout. Also known as "bush," or "bonfire."

CEILING—In general, the total distance from ground or water vertically to the base of lowest cloud layer that, in summation with all lower layers of clouds and obscuring phenomena, covers six-tenths or more of the sky. For detailed variations of the term, refer to the WBAN *Manual of Surface Observations,* "Circular N," U.S. Weather Bureau, Washington, D.C.

CEILOMETER—An automatic, recording, cloud height indicator.

CELSIUS—A temperature scale identical to the centigrade scale, with 0° as the melting point of ice and 100° as the boiling point of water.

CENTER OF ACTION—Any one of several large areas of high and low barometric pressure, changing little in location, and persisting through a season or through the whole year; for example, the Iceland low, the Siberian winter high. Changes in the intensity and positions of these pressure systems are associated with widespread weather changes.

CENTIGRADE—A scale of temperature whereon 100° represents the boiling point of pure water at standard atmospheric pressure, and 0° represents the melting point of pure-water ice. For conversion of centigrade degrees to Fahrenheit degrees, see page 29.

CERAUNOGRAPH—An electronic instrument for recording the occurrence of lightning discharges whether close by or so far away as to be invisible and their thunder unheard.

CHINOOK, or CHINOOK WIND—A Foehn blowing down the eastern slopes of the Rocky Mountains over the adjacent plains, in the United States and Canada. In winter, this warm, dry wind causes snow to disappear with remarkable rapidity, and hence it has been nicknamed the "snow-eater." The Santa Ana wind in Southern California is of the same type.

CHUBASCO—A violent squall on the west coast of tropical and subtropical North America.

CLIMATE—The prevalent or characteristic meteorological conditions of any place or region.

CLOUD—An accumulation of condensed water vapor, visible in the air. It may be composed of tiny water droplets in suspension, at low and medium altitudes, or of snow or fine ice crystals in the subfreezing temperatures of the upper altitudes.

CLOUD BANNER—A banner-like cloud streaming off from a mountain peak.

CLOUDBURST—A sudden, torrential downpour of rain usually from a thunderstorm (cumulonimbus) cloud.

CLOUD CAP—A cap-like cloud crowning either a mountain summit or another cloud, especially a mass of cumulonimbus.

CLOUDINESS (Sky Cover)—Amount of sky covered, but not necessarily hidden, by clouds or obscuring phenomena aloft or concealed by obscuring phenomena on the ground or both. (Circular "N," U.S. Weather Bureau.)

COL—An area on the weather map between two depressions and two eminences, giving a "saddleback" shape to that part of the map.

COLD AIR MASS—An air mass that is colder than the surface over which it is moving.

COLD DOME—A moving mountain or mass of cold, dense air.

COLD FRONT—A narrow strip or zone (indicated by a special line on the weather map) marking the boundary between two air masses and where cold air replaces warm air.

COLD WAVE—A rapid and marked fall of temperature during the cold season of the year. The U.S. Weather Bureau applies this term to a fall of temperature in 24 hours equaling or exceeding a specified number of degrees and reaching a specified minimum temperature or lower, the specifications varying for different parts of the country and for different periods of the year.

CONDENSATION—The process whereby water vapor is re-formed into water, the reverse of the process of evaporation. Condensation level is the height at which a rising column of air reaches saturation and clouds form.

CONDUCTION—Transfer of heat by molecular action.

CONSTANT-PRESSURE CHART—A chart which usually contains plotted data and analysis of the distribution of heights of any selected isobaric surface, and the analyses of the wind, temperature, and humidity existing at each height of the selected isobaric surface.

CONTINENTAL CLIMATE—The type of climate characteristic of the interior of a continent. As compared with a marine climate, a continental climate has a large annual and daily range of temperature.

CONTRAIL—A cloud-like streamer frequently observed to form behind aircraft flying in clear, cold, humid air.

CONVECTION—The process of transfer of heat in the atmosphere by ascending and descending currents, due to thermal instability.

CONVECTIONAL PRECIPITATION—Precipitation from clouds caused by thermal instability, that is, clouds of the cumulus or cumulonimbus type.

CONVERGENCE—An inflowing of air into a region from more than one direction in such manner that more air flows in than flows out. So long as this continues it will tend to produce an increasing pressure and temperature with their consequent effects. If the air is sufficiently humid, some of it, if forced upward, may cause a "convergent fog" or low stratus clouds.

CORIOLIS FORCE—A deflective force caused by the rotation of the earth on its axis. It causes moving bodies to be apparently deflected to the right of their course in the Northern Hemisphere and to the left in the Southern Hemisphere. The magnitude of this deflective force is basically dependent upon the velocity of the moving body and the latitude. (See also FERREL'S LAW.)

CUMULIFORM—A general term applied to all clouds having dome-shaped upper surfaces which exhibit protuberances, the bases of such clouds being generally horizontal. Cumuliform clouds are characteristically distinct and separated from one another by clear spaces.

CYCLOGENESIS—The term applied to the process which creates or develops a new cyclone. The word is applied also to the process which produces an intensification of a preexisting cyclone.

CYCLONE—This term, indicating in general a region of low atmospheric pressure, is now generally reserved to refer to violent disturbances originating in the tropics. In temperate and higher latitudes an area of low pressure is now usually designated as a *low* or a *depression*.

CYCLOSTROPHIC WIND—A wind which blows as a result of pressure gradient and centrifugal force, but in the absence of Coriolis force. It is, of necessity, cyclonic and restricted to the equatorial zones—the only place where Coriolis force is zero, or nearly so. Hurricanes are largely cyclostrophic until reaching latitudes high enough to be affected by Coriolis force.

DEEPENING—The occurrence of decreasing pressure in the center of a moving pressure system.

DENSITY—The mass of a unit volume of any substance.

DEPRESSION—See CYCLONE.

DEVIATION OF THE WIND—The angle between the direction of the wind and the direction of the pressure gradient. (See also INCLINATION OF THE WIND.)

DEVIL—The name applied to a dust whirlwind in India. The term is also current in South Africa. (In Western U.S. the term is *dust devil*.)

DEW—Condensed water vapor deposited on the surface of terrestrial objects.

DEW POINT—The temperature to which air must be cooled in order to become saturated, cooling below such temperature resulting in condensation.

DISCONTINUITY—The term applied in a special sense by meteorologists to a zone within which there is a comparatively rapid transition of the meteorological elements, particularly the boundary surface separating air masses of different temperatures.

DISTURBANCE—A local departure from the normal or average wind condition of any part of the world or, in other words, a feature of what is sometimes called the "secondary" circulation of the atmosphere, as distinguished from the general circulation. In everyday usage disturbance has come to be synonymous with *cyclone* and *depression*.

DIURNAL—Daily.

DIVERGENCE—The opposite of convergence.

DOLDRUMS—The tropical regions of the Equatorial Belt characterized by calms and light, shifting winds with frequent thunderstorms, squalls, and heavy showers.

DRIZZLE—Precipitation consisting of numerous tiny droplets. Drizzle originates from stratus clouds. The water droplets are so numerous the air seems full of them. "Drizzle" is essentially the same as "mist" in the U.S. weather code.

DROUGHT—A protracted period of dry weather.

DRY ADIABATIC LAPSE RATE—A rate of decrease of temperature with height approximately equal to 1°C. per 100 meters (1°F. per 180 ft). This is close to the rate at which an ascending body of unsaturated air will cool due to adiabatic expansion.

DRY BULB—A name given to an ordinary thermometer used to determine the temperature of the air, in order to distinguish it from the wet bulb.

DRY FOG—A haze due to the presence of dust or smoke in the air.

DYNAMIC COOLING—Decrease in temperature of air when caused by its adiabatic expansion in moving to a higher altitude or other region of lower pressure.

DYNAMIC METEOROLOGY—The branch of meteorology that treats of the motion of the atmosphere and its relations to other meteorological phenomena.

EDDY—A swirl in the air caused by flow over rough terrain or around terrestrial obstacles, or meeting with other air having different characteristics of density or motion.

EMINENCE—A region where the atmospheric pressure is higher than that of other surrounding regions. It is characterized by generally fair weather with a wind system of counterclockwise rotation. In the Northern Hemisphere, rains, if existent therein, are usually found in the southern or southeastern sector. It is the counterpart of a *depression* or region of *low* pressure.

EQUATORIAL AIR—Air of an air mass whose source lies in the Equatorial Belt or *doldrums*.

EQUIVALENT POTENTIAL TEMPERATURE—The temperature that a given sample of air would have if it were brought adiabatically to the top of the atmosphere (that is, to zero pressure) so that along its route all the water vapor present were condensed and precipitated, the latent heat of condensation being given to the sample, and then the remaining dry air compressed adiabatically to a pressure of 1000 mb. The equivalent potential temperature at any point is therefore determined by the values of absolute temperature, pressure, and humidity. It is one of the most conservative of air mass properties.

EVAPORATION—The transformation of water into water vapor, but the term can also be applied to most liquids. Evaporation is a cooling process.

EVAPORIMETER—An instrument for measuring the rate of evaporation of water into the atmosphere.

EYE OF THE STORM—A calm region at the center of a tropical cyclone or a break in the clouds marking its location.

FAHRENHEIT—A scale of temperature whereon 212° represents the boiling point of pure water at standard atmospheric pressure, and 32° represents the melting point of pure-water ice. For conversion of Fahrenheit degrees to centigrade degrees, see page 29.

FALL-WIND—A wind blowing down a mountainside; or any wind having a strong downward component. Fall-winds include the Foehn, mistral, bora, etc.

FALSE CIRRUS—Cirrus-like clouds at or somewhat below the summit of a thunder cloud; more appropriately called *thunder cirrus*.

FERREL'S LAW—"When a mass of air starts to move over the earth's surface, it is deflected to the right in the Northern Hemisphere, and to the left in the Southern Hemisphere, and tends to move in a circle whose radius depends upon its velocity and its distance from the equator."

FESTOON CLOUD—Mammatocumulus.

FILLING—The occurrence of increasing pressure in the center of a moving pressure system. *Filling* is the opposite of *deepening*.

FOEHN (or Föhn) WIND—A relatively warm, dry wind blowing along generally downward slopes. The air is heated in adiabatic compression as it flows to lower levels. Such winds are frequently caused by the subsidence of heavy, superior air from an upper air mass. In the United States they are found most frequently in the northwestern states and along the eastward slopes of the Rocky Mountain and Great Plains region.

FOG—Fog is formed when the air near the earth's surface is cooled below its dew point. Fog is, therefore, nothing but a cloud that touches the ground. By international agreement such a cloud is called *fog* when the visibility is less than one kilometer; if the visibility is greater than this, it is called *mist*, or *thin fog*.

FRICTION LAYER—The lower layer of air, below the "free atmosphere," where friction with the earth's surface affects its flow. Depending upon conditions, its thickness is usually from 1500 to 3000 ft.

FRONT—A definite boundary or mixing zone (a few miles wide) which occurs between two dissimilar air masses, or a surface of discontinuity between two juxtaposed currents of air possessing different densities, or, more simply, the boundary between two different air masses which have been brought together. The two air masses do not immediately mix and the denser, colder air always underlies or underrides the rarer (less dense) warm air in the form of a wedge. This condition of warm air overlying the colder air in the boundary or mixing zone, which is really the front, is sometimes called a *discontinuity surface* and is generally inclined at a slight angle to the surface of the earth.

FRONTOGENESIS—The term used to describe the process which creates a front, that is, produces a discontinuity in a continuous field of the meteorological elements; also applied to the process which increases the intensity of a preexisting front. Frontogenesis is generally set up by the horizontal convergence of air currents possessing widely different properties.

FRONTOLYSIS—The term used to describe the process which tends to destroy a preexisting front. Frontolysis is generally brought about by horizontal mixing and divergence of the air within the frontal zone.

FROST—A crystalline ice formation deposited on objects when the temperature is below 32°F. Sometimes caused by direct sublimation of the water vapor into ice crystals on an object whose temperature is below 32°F. Will also form sometimes on an airplane when it passes from a cold layer of air into a stratum of air of much higher temperature and moist content. It is dangerous to attempt a takeoff when the airplane is covered with frost as this greatly increases the drag by destroying the boundary layer of air next to the airfoil.

FROST SMOKE—See ARCTIC SMOKE.

GALE—Wind with an hourly velocity exceeding some specified value. In American practice a wind of force 7 to 10 on the Beaufort scale is counted a gale.

GENERAL CIRCULATION—The grand wind system of the entire earth, also called *atmospheric circulation*. (See Figure 6–12.)

GEOSTROPHIC WIND—The velocity of the wind in the free atmosphere attained when blowing under the conditions of complete balance of forces, that is, the "pressure force" due to the pressure gradient and the "anti-pressure component" of the Coriolis force. (See Figure 6–4.)

GLAZE (or Glazed Frost)—A coating of clear ice formed on terrestrial objects by rain which freezes upon contact.

GLORY—A series of concentric colored rings around the shadow of the observer, or of his head only, cast upon a cloud or fog bank. It is due to the diffraction of reflected light.

GRADIENT—The increase or decrease of a meteorological element with respect to distance, either horizontal or vertical; for example, a *barometric gradient* of plus 10 mb per thousand miles, or a *vertical temperature gradient* of −4°C. per thousand feet altitude, the latter being more commonly called the *lapse rate*. Temperature lapse rate is an indicator of atmospheric stability, while barometric gradient is an indicator of wind velocity. A steep barometric gradient indicates a rapid pressure drop with strong winds; a shallow one, the reverse.

GRADIENT WIND—The wind velocity necessary to balance the pressure gradient. The true wind above the friction layer is approximately equal to the gradient wind.

GRANULAR SNOW—Precipitation from stratus clouds (frozen drizzle) of small (1 mm or less in diameter) opaque grains of snow. Granular snow offers no appreciable icing conditions to aircraft in flight.

GREAT CIRCLE—A circle on the earth's surface, having the earth's diameter as its own diameter. A great circle can be drawn between any two points on the globe, and the arc between them is the shortest distance between the two points.

GUST—A sharp increase in wind velocity of short duration.

HAIL—Hard ice globules falling from the clouds. *Soft hail* is milky white, pithy, and soft.

HALO—A generic name for a large group of optical phenomena caused by ice crystals in the atmosphere. The commonest of these phenomena is the halo of 22° (that is, of 22° radius), surrounding the sun or moon. The halo of 46° and the rare halo of 90°, or halo of Hevelius, also surround the luminary. Other forms of halo are the tangent arcs, parhelia (or paraselenae), parhelic (or paraselenic), circle, anthelion, etc.

HAZE—Very fine particles of moisture, smoke, and dust suspended in the air. It decreases visibility.

HIGH—See EMINENCE.

HORSE LATITUDES—The regions of calms in the subtropical anticyclone belts.

HOT WAVE—A period of abnormally high temperatures. It has sometimes been defined, in the United States, as a period of three or more consecutive days during each of which the temperature is 90°F. or over.

HUMIDITY—Water-vapor content of the air. (See also RELATIVE and SPECIFIC HUMIDITY.)

HURRICANE—Violent disturbance of tropical origin (see CYCLONE), any wind above 75 mph.

HYDROMETEOR—A generic term for weather phenomena such as rain, cloud, fog, etc., which depend mostly upon modification in the condition of the water vapor in the atmosphere.

HYGROGRAPH—An instrument which makes a continuous record of humidity.

HYGROMETER—An indicator which indicates the momentary value of the humidity.

HYGROTHERMOGRAPH—An instrument which makes a continuous record of both humidity and temperature.

ICEBERG—A large mass of ice that breaks from the tongue of a glacier running into the sea and floats away.

ICE NEEDLES—Thin crystals or shafts of ice, so light that they seem to be suspended in the air.

INCLINATION OF THE WIND—The angle which the wind direction makes with the direction of the isobar at the place of observation. Over the ocean the angle is approximately 10°, while over land it is usually between 20° and 30°. (See also DEVIATION OF THE WIND.)

INSOLATION—Solar radiation, as received by the earth or other planets; also, the rate of delivery of the same, per unit of horizontal surface.

INSTABILITY—That condition of saturated air when its lapse rate is greater than the moist adiabatic, or of unsaturated air when its lapse rate is greater than the dry adiabatic. Instability is evidenced in ascending thermal currents, cumulus-type clouds, and thunderstorms.

INTERTROPICAL FRONT—The boundary between the trade wind system of the Northern and Southern Hemispheres. It manifests itself as a fairly broad zone of transition commonly known as the *doldrums*.

INVERSION—A stratum in the air where the lapse rate is such that the temperature increases (instead of decreases) with altitude.

IONOSPHERE—A layer of ionized air high above the earth's surface. It actually appears to be composed of at least three layers, E, F_1 and F_2, at heights varying from 40 or 50 to 175 or 200 miles.

ISALLOBARS—On a weather map, lines of equal barometric tendency.

ISOBARS—On a weather map, lines of equal barometric pressure.

ISOTACH—A line of equal or constant wind speed.

ISOTHERMS—On a weather map, lines of equal temperature.

JET STREAM—A meandering river of high-velocity winds, 50 knots or greater, imbedded in the normal wind flow aloft, often 1000 to 3000 miles long, 100 to 400 miles wide. The core of the jet stream is generally found at altitudes of 20,000 to 40,000 ft. Several individual jet streams have been isolated, so caution must be exercised as different people sometimes define the jet stream differently.

KNOT—A knot is a speed equal to one nautical mile (6080 ft) per hour. it is the standard unit of speed for marine and air navigation. (Its reference is to speed or velocity only, and is not used as a measure of distance.)

LAND AND SEA BREEZES—The breezes that, on certain coasts and under certain conditions, blow from the land by night and from the water by day.

LAPSE RATE—The rate of decrease of temperature in the atmosphere with height; usually the decrease in temperature with altitude is 3.5°F. for each 1,000 ft. (See also ADIABATIC PROCESS.)

LENTICULAR CLOUD—A cloud having approximately the form of a double-convex lens. Clouds of this sort may be formed at the crests of standing waves of atmosphere such as are often induced by mountain ranges; usually they represent a traditional stage in the development or disintegration of one of the more well-known cloud types.

LID—A term often used to denote a temperature inversion in the atmosphere. Since the air in an inversion is stable, convectional currents cannot exist within it. For this reason an inversion is called a *lid*, since it prevents the air below and above it from mixing. Lids may vary in thickness from a few feet to several hundred feet, according to the conditions.

LIGHTNING—The neutralizing electric flash between two oppositely charged areas.

LINE SQUALL—A long cumulonimbus cloud (*cumulonimbus arcus*) extending along a cold front, with violent winds, rain, thunder and lightning, and, frequently, heavy hail. With a strong cold front, these squalls may extend from 100 to 400 miles in an almost unbroken line making it impracticable to fly around them. They may extend to such altitudes as to make it impracticable to fly over them, and the violent winds and torrential rain below them make it dangerous to attempt to fly in the area underneath them where may be found heavy downdrafts extending all the way to the ground. (See Figures 9–14A and B.)

LOW—See CYCLONE.

MACKEREL SKY—A sky covered with or containing large patches of cirrocumulus or altocumulus clouds resembling the scales and stripes on the side of a mackerel.

MAMMATOCUMULUS—A form of cloud showing pendulous, sac-like protuberances.

MARINE CLIMATE—A type of climate characteristic of the ocean and oceanic islands. Its most prominent feature is equability of temperature.

MAXIMUM—The highest value of any element occurring during a given period.

MENISCUS—The curved upper surface of a liquid in a tube.

METEOROLOGY—The science of the atmosphere.

METEOROGRAPH—A self-recording instrument for recording pressure, temperature, and humidity in the free air.

MICROBAROGRAPH—An instrument designed for recording small and rapid variations of atmospheric pressure.

MILLIBAR—See BAR.

MINIMUM—The lowest value of any element occurring during a given period.

MIRAGE—An apparent displacement or distortion of observed objects by abnormal atmospheric refraction. Sometimes the images of objects are inverted, magnified, multiplied, raised, or brought nearer to the eye than the object. Refraction layers in the atmosphere often assume the appearance of fog.

MIST—A very thin fog, in which the horizontal visibility is greater than 1 kilometer, or approximately 1100 yards. (This is the definition laid down by the International Meteorological Organization.) In North America the word is often used synonymously with drizzle or fine rain.

MONSOON—Seasonal winds which blow with great steadiness, reversing their direction with the change of season. In summer their direction is generally toward the large heated land areas, rushing in to displace the great volumes of air rising in convective currents. In winter they blow, with less force, outward from the great areas of ice and snow toward the tropics or out over the surface of the warmer ocean.

NEPHOSCOPE—An instrument used in observing the direction and velocity of cloud movement.

NEUTRAL POINT—The term applied in a special sense to any point at which the axis of a wedge of high pressure intersects the axis of a trough of low pressure. Also called saddle point. (See also COL.)

NOCTILUCENT CLOUDS—Luminous, cirrus-like clouds sometimes visible throughout the short nights of summer; supposed to be clouds of dust at great altitudes shining with reflected sunlight. Such clouds were observed during several summers after the eruption of Krakatoa (1883) and are still occasionally reported.

NUCLEUS—A particle upon which condensation of water vapor occurs in the free atmosphere in the form of a water drop or an ice crystal.

OBSCURATION (PARTIAL)—The situation of sky cover when a surface-based phenomenon obscures a portion of the sky from the point of observation.

OBSCURATION (TOTAL)—The situation of sky cover when a surface-based phenomenon obscures all of the sky from the point of observation. The number preceding the X symbol indicates vertical visibility into the obscuring phenomenon, in hundreds of feet. Example: W3X$\frac{1}{4}$F, the vertical visibility is 300 ft. (The horizontal visibility is $\frac{1}{4}$ statute mile.)

OCCLUDED FRONT—See OCCLUSION.

OCCLUSION—The zone in which, when one front overtakes another, it forces one front upward from the surface of the earth. The front is said to be an *occluded front*.

OROGRAPHIC LIFTING—The lifting of air caused by its flow up the slopes of hills or mountains.

OROGRAPHIC RAIN—Rain resulting from orographic lifting.

OROGRAPHY—The science which treats of mountains.

OZONE—A colorless gas (O_3) obtained as an allotropic form of oxygen. It is a strong oxidizer, and has a chlorine-like odor. It has an important absorbing effect on ultraviolet radiation. It exists in rather slight quantity near the earth's surface, increasing to a maximum at an average altitude of about 35,000 ft. It can often be smelled in the air near the surface immediately following a thunderstorm, particularly in the immediate vicinity of lightning flashes.

PARHELION (plural PARHELIA)—A mock sun, or sun dog; a form of halo consisting of a more or less distinctly colored image of the sun at the same altitude as the latter above the horizon, and hence lying on the parhelic circle, if present. The ordinary parhelia are 22° from the sun in azimuth, or a little more, according to the altitude of the luminary. Parhelia have occasionally been seen about 46° from the sun. Analogous phenomena seen in connection with the moon are called paraselenae, mock moons, or moon dogs. (See also HALO.)

PILOT BALLOON—A small balloon inflated with hydrogen or helium until the buoyancy is sufficient to float the balloon in air with prescribed counterweights attached, thus giving it a known ascensional rate. When cast adrift it is employed in determining the direction and velocity of the wind at various levels.

POLAR AIR—Air of an air mass whose source lies in the region of the poles.

POLAR CONTINENTAL AIR—The term used to describe any air mass that originates over land or frozen ocean areas in the polar regions. Polar continental air is characterized by low temperatures, low specific humidity, and a high degree of vertical stability.

POLAR FRONT—The general frontal zone where air from the cold polar regions meets the warm tropical air from the subtropical anticyclone belts; more clearly defined over oceans than over continents.

POLAR MARITIME AIR—The term used to describe any air mass that originally came from the polar regions but has since been modified by reasons of its passage over a relative warm ocean surface. Polar maritime air is characterized by moderately low surface temperatures, moderately high surface specific humidity, and a considerable degree of vertical instability.

POTENTIAL TEMPERATURE—The resultant temperature attained by a sample of air when compressed adiabatically to standard pressure of 1000 mb (that is, approximately sea level).

PRECIPITATION—The product of condensation of water vapor in the atmosphere.

PRESSURE—An elliptical expression, current in meteorological literature for atmospheric pressure, or barometric pressure.

PRESSURE ALTITUDE—The altitude indicated by an altimeter when the barometric scale thereof is adjusted to the standard sea level atmospheric pressure of 29.92 in. of mercury.

PRESSURE GRADIENT—The rate of change in barometric pressure per unit horizontal distance in the direction in which the pressure changes most rapidly.

PRESSURE TENDENCY—Same as BAROMETRIC TENDENCY.

PREVAILING VISIBILITY—The greatest horizontal visibility which is equaled or surpassed throughout half of the horizon circle; it need not be a continuous half.

PREVAILING WESTERLIES—Winds which blow toward the poles from the horse latitudes. Not so steady as the trade winds, but more constant in the Southern Hemisphere, due to oceans and lack of land interruptions. They are known as the "roaring forties" owing to the latitude at which they occur.

PREVAILING WIND—The wind direction most frequently observed during a given period.

PSEUDO-ADIABATIC PROCESS—The process of saturated air rising in the atmosphere and losing its water vapor by condensation and precipitation.

PSYCHROMETER—An instrument for measuring humidity. It is composed of two thermometers one of which has a water-saturated wick covering the thermometer bulb, called the *wet bulb*. Evaporation from the wick cools the wet bulb, making it read lower than the other (*dry bulb*) thermometer. The drier (less humid) the air, the more rapid the evaporation and, consequently, the greater the difference in temperature between the wet and dry bulbs. With this difference in temperature and the dry bulb temperature as arguments, the relative humidity is obtained from a Humidity Table published by the U.S. Weather Bureau. When the air is saturated, that is, 100% relative humidity, evaporation ceases and the wet and dry bulbs read exactly the same. In the "sling" type psychrometer (Figure 5–3) the instrument is whirled around to produce the evaporation from the wet bulb.

PUMPING—Unsteadiness of the mercury in the barometer caused by fluctuations of the air pressure produced by a gusty wind, or due to the oscillation of a ship.

PYRHELIOMETER—An instrument that measures solar radiation by its heating effects.

QUASI-STATIONARY FRONT—The ideal stationary front is seldom found in nature, but it often occurs that the frontal movement is such that no appreciable displacement takes place. The front is then said to be quasi-stationary. Such a front constitutes the most intricate forecasting problem, for (1) as the velocity of the front is nil, its future movement will depend on the acceleration, which is exceedingly difficult to evaluate or even estimate, and (2) the lack of movement of the front is highly favorable for the formation of cyclonic disturbances.

RADIATION—Emission of energy in the form of waves, either light or heat or both. Travel of electromagnetic waves at 186,000 miles per second, many of which may be visible as light. Cosmic rays, gamma rays, X-rays, ultraviolet rays, visible light rays, infrared rays, and radio waves are some common types of radiation.

RADIATION FOG—Fog characteristically resulting from the radiational cooling of air near the surface of the ground on calm, clear nights.

RADIOMETEOROGRAPH—Same as RADIOSONDE.

RADIOSONDE—A small pilot balloon which carries a set of measuring instruments aloft; a small parachute lowers it to earth again when the balloon bursts in the upper atmosphere. By means of a small clockwork motor and very lightweight radio transmitting set, the indications of instruments sensitive to pressure, temperature and humidity are automatically transmitted at regular intervals during the flight. The signals from the radiosonde are received and recorded on a special receiver on the ground and are then translated into readings of pressure, temperature, and humidity at the various altitudes. Rewards are paid to persons who find and return radiosondes to the Weather Bureau.

RAIN—Water drops in the atmosphere, the result of precipitation.

RAINBOW—A luminous arc formed by the refraction and reflection of light in drops of water.

RAINFALL—The term sometimes synonymous with rain, but most frequently used in reference to amounts of precipitation (including snow, hail, etc.)

RAIN GAUGE—An instrument for measuring the amount of rainfall on the earth's surface.

REFRACTION—The bending of light rays when they pass from a medium of one density into or through a medium of a different density.

RELATIVE HUMIDITY—The ratio of the actual moisture content of the air to the amount of moisture that the air could hold, saturated, at the same temperature. This is expressed as percentage.

REPRESENTATIVE OBSERVATIONS—Those which give the true or typical meteorological conditions prevailing in an air mass; hence they must be relatively uninfluenced by local conditions.

RHUMB LINE—A line on the earth's surface which cuts all meridians at the same angle. Thus an aircraft flying a steady, true course is following a Rhumb line, or *loxodromic curve.*

RIDGE—An elongated area of high pressure extending from an eminence, also called a wedge. (See Figure 4–5.)

RIME ICE—White, opaque, granular ice which forms on various parts of aircraft in flight under certain conditions.

ROTATION OF THE EARTH (effect of)—See CORIOLIS FORCE.

SADDLE—See COL.

ST. ELMO'S FIRE—A luminous brush discharge of electricity which appears on sharp points or edges of objects during the existence of strong electrical fields. It may appear as a general glow on the object or as numerous streamers. It is often seen on the wing tips and propellers of aircraft flying in or near thunder clouds.

SATURATED ADIABATIC LAPSE RATE—See ADIABATIC PROCESS (Moist-adiabatic lapse rate).

SATURATION—The condition that exists in the atmosphere when the partial pressure exerted by the water vapor present is equal to the maximum vapor pressure possible at the prevailing temperature. The condition of air which contains all the moisture it can hold.

SCARF CLOUD—A thin cirrus-like cloud which often drapes the upper parts of tall cumulonimbus clouds.

SCUD—Low cloud fragments drifting below the main cloud layers, particularly following or during rainfall.

SEA BREEZES—See LAND AND SEA BREEZES.

SECONDARY FRONT—A second front of similar nature to and following fairly closely behind a primary front. A disturbance connected therewith is called a *secondary disturbance.* Secondary disturbances frequently develop into much worse weather than, and have the effect of "killing," the primary disturbance.

SEMICIRCLE—The "dangerous semicircle" of a cyclone storm at sea is the half of the storm area in which rotary and progressive motions of the storm reinforce each other, and the winds are also directed in such a way as to drive a vessel running before the wind across the storm track ahead of the advancing center. The other half is called the "navigable" semicircle.

SHOWER—Precipitation of a rapidly varying intensity, falling from convective (cumulus type) clouds.

SKY COVER—See CLOUDINESS.

SLEET—In the United States, sleet is another name for "freezing rain," that is, rain which freezes on striking terrestrial objects. In Europe, sleet is defined as rain and snow falling simultaneously.

SMOG—A mixture of smoke and fog.

SMOKE—The products of incomplete combustion, suspended in the air.

SNOW—Condensed water vapor, congealed into white or transparent crystals or flakes in the air.

SOFT HAIL—White, opaque, round pellets of snow.

SOUNDING—An investigation of meteorological conditions at various altitudes above the earth's surface. Air soundings may be made by means of pilot (sounding) balloons for obtaining winds aloft, or by aerometeorgraph or radiosonde for obtaining pressure, temperature, and humidity.

SOUNDING BALLOON—See PILOT BALLOON.

SOURCE REGION—An extensive area of the earth's surface characterized by essentially uniform surface conditions and so placed in respect to the general atmospheric circulation that air masses may remain over it long enough to acquire definite characteristic properties. Examples of source regions are the ice-covered polar regions and the broad expanses of uniformly warm tropical oceans.

SPECIFIC HUMIDITY—The number of grams of water vapor contained in one kilogram of air.

SQUALL—A sudden and violent wind often attended with rain or snow. A squall lasts for a matter of minutes as compared to a *gust* which usually passes in a matter of seconds.

SQUALL LINE—Any nonfrontal line or narrow band of active thunderstorms (with or without squalls); a mature instability line.

STABILITY—The state air is in if, when displaced from its original level, it tends to return thereto. This holds true for saturated air when its lapse rate is less than the moist adiabatic, and for unsaturated air when its lapse rate is less than the dry adiabatic. (See also INSTABILITY.)

STAGNATION POINT—The point on the leading edge of a body in the moving air stream which is the division point for the lines of air flow on either side of the body. The air is practically stationary at this point, and static pressure prevails, according to Bernoulli's theory.

STANDARD ATMOSPHERE—A standard atmosphere is defined as one in which the sea-level temperature is 15°C.; the sea-level pressure, 1013.2 mb.; and the lapse rate, 6.5°C. per kilometer up to 11 km (the stratosphere). In the foot-pound-second system this amounts to very nearly a sea-level temperature of 59°F., a sea-level pressure of 29.92 in. of mercury, and a lapse rate of nearly 3.5°F. per 1000 ft of altitude. The real atmosphere duplicates the standard atmosphere only occasionally.

STATIC—In meteorology, a state in which the position or properties of an air mass or frontal zone are not changing or not moving, such as a "static front" or "static conditions." (See also STATIONARY FRONT, QUASI-STATIONARY FRONT.)

STATIONARY FRONT—A front along which one air mass does not replace the other.

STORM—A violent disturbance of the atmosphere, either by wind or by other undesirable meteorological conditions, such as rain, thunder and lightning, dust or sand, ice or sleet.

STRATIFORM—A general term applied to all clouds which are arranged in unbroken horizontal layers or sheets.

STRATOSPHERE—The upper portion of the atmosphere lying above the tropopause. Motion of the air therein is mainly in horizontal, stratified flow with almost complete absence of vertical currents. (See Figure 2–1.)

SUBLIMATION—At low temperatures the condensation nuclei become inactive and the sublimation nuclei become active. Under these conditions, water vapor present in the air changes directly from the vapor state to the solid state. An example of the sublimation process is the formation of frost, a condition wherein the vapor in the atmosphere sublimates, or changes directly into ice crystals without passing through the liquid state.

SUBSIDENCE—A slow, downward "settling" of air from upper altitudes.

SUNSHINE RECORDER—An instrument for recording the duration of sunshine; certain types also record the intensity of sunshine.

SUPERIOR AIR—Believed to be developed by strong subsiding motions. Most frequently observed at higher and intermediate levels over the southwestern United States, it is found sometimes at the surface particularly in the South Central States. When aloft it is cold and dry, but when brought down by general subsidence, it is heated adiabatically in descent and produces the warm, dry Chinook or Foehn winds flowing down mountain sides and long downward slopes. Flying conditions in superior air are excellent.

SURGE—A general change in barometric pressure apparently superposed upon cyclonic and normal diurnal changes.

SYNOPTIC CHART—Technical name for a weather map, inasmuch as it depicts a synopsis of meteorological conditions over a large area at a given moment.

SYNOPTIC METEOROLOGY—The branch of meteorology that deals with the analysis of meteorological observations made simultaneously at a number of points in the atmosphere (at the ground or aloft) over the whole or a part of the earth, and the application of the analysis to weather forecasting and other problems. ("Synoptic" comes from two Greek words meaning "general view.")

TEMPERATURE—A measure of the degree of hotness or coldness of a substance.

THERMOGRAM—The continuous record of temperature made by a thermograph.

THERMOGRAPH—An instrument used to make a continuous record of temperature.

THERMOMETER—An instrument used to measure temperature.

THUNDER—The sound folowing a discharge of lightning.

THUNDERSTORM—A storm attended by thunder and lightning. Thunderstorms are local disturbances, often occurring as episodes of cyclones, and, in common with squalls, are marked by abrupt variations in pressure, temperature, and wind.

TOPOGRAPHY—Descriptive features of any given portion of the earth's surface.

TORNADO—An extremely violent, rapidly whirling storm of great intensity, small diameter, and steep pressure gradients. It is easily recognized by its dark funnel-shaped cloud. The most violent of all storms, it is highly destructive but fortunately short-lived.

TRADE WINDS—Steady seasonal winds which blow from the subtropical anticyclone belts toward the belt of lower pressure in the equatorial regions. They blow from the northeast in the Northern Hemisphere and from the southeast in the Southern Hemisphere.

TRAJECTORY—The path traced out by a small volume of air in its movement over the earth's surface.

TRANSITION ZONE—The relatively narrow region occupied by a front wherein the meteorological properties exhibit large variations over a short distance and possess values intermediate between those characteristic of the air masses on either side of the zone.

TRIPLE INDICATOR—In weather map analysis, a "discontinuity" or marked change in *temperature, dew point,* and *wind direction* across a relatively narrow area on a synoptic chart indicating the existence of a frontal zone in that area. A change of two of the three items may or may not indicate a frontal zone, but a simultaneous change in all three is an almost positive indication. Similarly, when such a change occurs in a short period of time on the instruments at a weather station, it is an almost sure indication that a front has just passed the station.

TROPICAL AIR—Air of an air mass whose source lies in the equatorial belt or the subtropical eminences.

TROPICAL CYCLONE—Violent, whirling storm of destructive intensity originating in the tropics as small, concentrated and deep depressions.

TROPICAL DISTURBANCE—The name used by the Weather Bureau for a cyclone wind system of the tropics that is not known to have sufficient force to justify the use of the words "storm" or "hurricane."

TROPICAL MARITIME AIR—The term used to describe any air mass that originates over an ocean area in the tropics. Tropical maritime air is characterized by high surface temperatures and high specific humidity.

TROPOPAUSE—The boundary between the troposphere and the stratosphere.

TROPOSPHERE—All the earth's atmosphere lying between the tropopause and the surface. All "weather" occurs in the troposphere, with its variable winds, currents, clouds, storms, and other phenomena, whereas stratosphere conditions are clear, cloudless, steady, and stable.

TROUGH—An elongated area of low pressure extending from a depression.

TURBULENCE—Irregular motion of the atmosphere produced when air flows over a comparatively uneven surface, such as the surface of the earth, or when two currents of air flow past or over each other in different directions or at different speeds. The existence of turbulence in the atmosphere is made apparent by the character of the trail of smoke from a ship's funnel and by gusts and lulls in the wind.

TWILIGHT—Astronomical twilight is the interval between sunrise or sunset and the total darkness of night. Civil twilight is the period of time before sunrise and after sunset during which there is enough daylight for ordinary outdoor occupations.

TYPHOON—A tropical cyclone common to the China Sea and caused almost entirely by convection.

U-SHAPED DEPRESSION—A trough with U-shaped isobars, in which fronts rarely form.

VANE—A device that shows which way the wind blows; also called *weather vane* or *wind vane*.

VAPOR PRESSURE—The partial pressure of the water vapor contained in the atmosphere.

V-SHAPED DEPRESSION—A trough with V-shaped isobars, in which fronts frequently exist or form.

VEERING—The reverse of *backing*. A shift to the right (that is clockwise) in wind direction.

VERNIER—An auxiliary scale for estimating fractions of a scale division when the reading to the nearest whole division on the main scale is not sufficiently accurate.

VIRGA—Precipitation falling from clouds, but evaporating before hitting the earth's surface.

VISIBILITY—The horizontal transparency of the atmosphere at the surface; the greatest distance at which an object can be recognized *as such* by the average unaided eye.

WARM AIR MASS—An air mass that is warmer than the surface over which it is moving.

WARM FRONT—A frontal surface (or zone) between two air masses wherein air of a lower temperature is being displaced by air of a higher temperature.

WARM SECTOR—That area in a depression in which relatively warm air is flowing toward the "center" of the disturbance. This warm air (usually of tropical origin) supplies the energy upon which the depression "feeds." When the supply of warm air is cut off by the cold front overtaking the warm front and forming an occlusion, the depression "dies."

WATERSPOUT—A disturbance at sea very similar to a tornado over the land, but usually less violent.

WAVE—An undulation along the boundary between a cold and a warm air mass which usually results in the formation of a depression with a warm and a cold front, that is, if the wave is of an *unstable* nature.

WEDGE—See RIDGE, and Figure 4–5.

WET BULB—See PSYCHROMETER.

WIND—Air in approximately horizontal motion. Streams of air moving vertically are called *air currents*.

WIND ROSE—(1) A diagram showing the relative frequency and sometimes also the average strength of the wind blowing from different directions in a specified region. (2) A diagram showing the average relation between winds from different directions and the occurrence of other meteorological phenomena.

ZONDA—A chinook-type wind of Argentina.

COLLOQUIAL TERMS

BAGUIO—The name current in the Philippines for a tropical cyclone.

BORA—A cold wind of the northern Adriatic, blowing down from the high plateaus to the northward. Also, a similar wind on the northeastern coast of the Black Sea.

BRAVE WEST WINDS—The boisterous westerly winds blowing over the ocean between latitudes 40° and 50° south. This region is known as the *Roaring Forties*.

BULL'S-EYE—(1) A patch of clear sky at the center of a cyclonic storm; the "eye of the storm." (2) A small, isolated cloud seen at the beginning of a bull's-eye squall, marking the top of the otherwise invisible vortex of the storm.

BULL'S-EYE SQUALL—A squall forming in fair weather, characteristic of the ocean off the coast of South Africa; so called on account of the peculiar appearance of a small isolated cloud that marks the top of the invisible vortex of the storm.

CALLINA—A Spanish name for dry fog.

CAT'S PAW—A slight and local breeze which shows itself by rippling the surface of the sea.

CHUBASCO—A violent squall on the west coast of tropical and subtropical North America.

CORDONAZO; in full, CORDONAZO DE SAN FRANCISCO ("last of St. Francis")—A hurricane wind blowing from a southerly quadrant on the west coast of Mexico as the result of the passing of a tropical cyclone off the coast.

DEVIL—The name applied to a dust whirlwind in India. The term is also current in South Africa.

INDIAN SUMMER—The period of mild, calm, hazy weather occurring in autumn or early winter, especially in the United States and Canada; popularly regarded as a definite event in the calendar, but weather of this type is really of irregular and intermittent occurrence.

MACKEREL SKY—An area of sky covered with cirrocumulus clouds; especially when the clouds resemble the pattern seen on the backs of mackerel.

MARE'S TAILS—Cirrus in long slender streaks.

MISTRAL—Along the Mediterranean coast from the mouth of the Ebro to the Gulf of Genoa, a stormy cold northerly wind, blowing down from the mountains of the interior. (The name is sometimes applied to northerly winds on the Adriatic, in Greece, and in Algeria.)

MOUNTAIN AND VALLEY BREEZES—The breezes that in mountainous regions normally blow up the slopes by day (valley breezes) and down the slopes by night.

NORTHER—A northerly wind, especially a strong northerly wind of sudden onset, occurring during the colder half of the year over the region from Texas southward, including the Gulf of Mexico.

ROARING FORTIES—See BRAVE WEST WINDS.

SCUD—Shreds of small detached masses of cloud moving rapidly below a solid deck of higher clouds. Scud may be composed of either fractocumulus or fractostratus clouds.

SUN DOG—A mock sun or parhelion.

SUN DRAWING WATER—The sun is popularly said to be "drawing water" when crepuscular rays extend down from it toward the horizon. The sun's rays, passing through interstices in the cloud, are made visible through illumination of particles of dust in the atmosphere along their paths.

TABLECLOTH—A sheet of cloud that sometimes spreads over the flat top of Table Mountain, near Cape Town.

TEHUANTEPECER—A strong to violent northerly wind over Pacific waters off southern Mexico and northern Central America, confined mostly to the Gulf of Tehuantepec, and occurring during the colder months.

WOOLPACK—Cumulus.

APPENDIX 2

FAA REGULATIONS RELATED TO WEATHER

In this appendix, the Federal Aviation Regulations related to weather will be cited, using the FAA numbered sections and extracting from the general operating and flight rules those sections which are pertinent to weather.

§91.5 Preflight action.

Each pilot in command shall, before beginning a flight, familiarize himself with all available information concerning that flight. This information must include, for a flight under IFR or a flight not in the vicinity of an airport, available weather reports and forecasts, fuel requirements, alternatives available if the planned flight cannot be completed, and any known traffic delays of which he has been advised by ATC.

§91.75 Compliance with ATC clearances and instructions.

(a) When an ATC clearance has been obtained, no pilot in command may deviate from that clearance, except in an emergency, unless he obtains an amended clearance. However, except in positive controlled airspace, this paragraph does not prohibit him from canceling an IFR flight plan if he is operating in VFR weather conditions.

(b) Except in an emergency, no person may, in an area in which air traffic control is exercised, operate an aircraft contrary to an ATC instruction.

(c) Each pilot in command who deviates, in an emergency, from an

ATC clearance or instruction shall notify ATC of that deviation as soon as possible.

(d) Each pilot in command who (though not deviating from a rule of this subpart) is given priority by ATC in an emergency, shall, if requested by ATC, submit a detailed report of that emergency within 48 hours to the chief of that ATC facility.

§91.81   Altimeter settings.

(a) Each person operating an aircraft shall maintain the cruising altitude or flight level of that aircraft, as the case may be, by reference to an altimeter that is set, when operating—

(1) Below 18,000 feet MSL, to—

   (i) the current reported altimeter setting of a station along the route and within 100 nautical miles of the aircraft;

   (ii) If there is no station within the area prescribed in subdivision (i) of this subparagraph, the current reported altimeter setting of an appropriate available station; or

   (iii) In the case of an aircraft not equipped with a radio, the elevation of the departure airport or an appropriate altimeter setting available before departure; or

(2) At or above 18,000 feet MSL, to 29.92" Hg.

(b) The lowest usable flight level is determined by the atmospheric pressure in the area of operation, as shown in the following table:

CURRENT ALTIMETER SETTING	LOWEST USABLE FLIGHT LEVEL
29.92 (or higher)	180
29.91 thru 29.42	185
29.41 thru 28.92	190
28.91 thru 28.42	195
28.41 thru 27.92	200
27.91 thru 27.42	205
27.41 thru 26.92	210

(c) To convert minimum altitude to the minimum flight level, the pilot shall take the flight-level equivalent of the minimum altitude in feet and add the appropriate number of feet specified below, according to the current reported altimeter setting:

CURRENT ALTIMETER SETTING	ADJUSTMENT FACTOR
29.92 (or higher)	None
29.91 thru 29.42	500'
29.41 thru 28.92	1000'
28.91 thru 28.42	1500'
28.41 thru 27.92	2000'
27.91 thru 27.42	2500'
27.41 thru 26.92	3000'

§91.83   Flight plan; information required.

(a) Unless otherwise authorized by ATC, each person filing an IFR or VFR flight plan shall include in it the following information:

(1) The aircraft identification number and, if necessary, its radio call sign.

(2) The type of the aircraft or, in the case of a formation flight, the type of each aircraft and the number of aircraft, in the formation.

(3) The full name and address of the pilot in command, or, in the case of a formation flight, the formation commander.

(4) The point and proposed time of departure.

(5) The proposed route, cruising altitude (or flight level), and true airspeed at that altitude.

(6) The point of first intended landing and the estimated elapsed time until over that point.

(7) The radio frequencies to be used.

(8) The amount of fuel on board (in hours).

(9) In the case of an IFR flight plan, an alternate airport, except as provided in paragraph (b) of this section.

(10) In the case of an international flight, the number of persons in the aircraft.

(11) Any other information the pilot in command or ATC believes is necessary for ATC purposes.

When a flight plan has been filed, the pilot in command, upon cancelling or completing the flight under the flight plan, shall notify the nearest FAA Flight Service Station or ATC facility.

(b) Paragraph (a) (9) of this section does not apply if Part 97 of this subchapter prescribes a standard instrument approach procedure for the first airport of intended landing and the weather conditions at that airport are forecast to be, from two hours before to two hours after the estimated time of arrival, a ceiling of at least 1,000 feet above the lowest MEA, MOCA, or altitude prescribed for the initial approach segment of the instrument approach procedure for the airport and visibility at least three miles, or two miles more than the lowest authorized landing minimum visibility, whichever is greater.

(c) *IFR alternate airport weather minimums.* Unless otherwise authorized by the Administrator, no person may include an alternate airport in an IFR flight plan unless current weather forecasts indicate that, at the estimated time of arrival at the alternate airport, the ceiling and visibility at that airport will be at or above the following alternate airport weather minimums:

(1) If an instrument approach procedure has been published in Part 97 for that airport, the alternate airport minimums specified in that procedure or, if none are so specified, the following minimums:

(i) Precision approach procedure: ceiling 600 feet and visibility 2 statute miles.

(ii) Non-precision approach procedure: ceiling 800 feet and visibility 2 statute miles.

(2) If no instrument approach procedure has been published in Part 97 for that airport, the ceiling and visibility minimums are those allowing descent from the MEA, approach, and landing, under basic VFR.

VISUAL FLIGHT RULES

§91.105 Basic VFR weather minimums.

(a) Except as provided in §91.107, no person may operate an aircraft under VFR when the flight visibility is less, or at a distance from clouds that is less, than that prescribed for the corresponding altitude in the following table:

ALTITUDE	FLIGHT VISIBILITY	DISTANCE FROM CLOUDS
1,200 feet or less above the surface (regardless of MSL altitude)—		
Within controlled airspace	3 statute miles	500 feet below. 1,000 feet above. 2,000 feet horizontal.
Outside controlled airspace	1 statute mile except as provided in § 91.105(b).	Clear of clouds.
More than 1,200 above the surface but less than 10,000 feet MSL—		
Within controlled airspace	3 statute miles	500 feet below. 1,000 feet above. 2,000 feet horizontal.
Outside controlled airspace	1 statute mile	500 feet below. 1,000 feet above. 2,000 feet horizontal.
More than 1,200 above the surface and at or above 10,000 feet MSL.	5 statute miles	1,000 feet below. 1,000 feet above. 1 mile horizontal.

(b) When the visibility is less than one mile, a helicopter may be operated outside controlled airspace at 1,200 feet or less above the surface if operated at a speed that allows the pilot adequate opportunity to see any air traffic or other obstruction in time to avoid a collision.

(c) Except as provided in §91.107, no person may operate an aircraft, under VFR, within a control zone beneath the ceiling when the ceiling is less than 1,000 ft.

(d) Except as provided in §91.107, no person may take off or land an aircraft, or enter the traffic pattern of an airport, under VFR, within a control zone—

(1) Unless ground visibility at that airport is at least three statute miles; or

(2) If ground visibility is not reported at that airport, unless flight visibility during landing or take off, or while operating in the traffic pattern, is at least three statute miles.

(e) For the purposes of this section, an aircraft operating at the base altitude of a transition area or control area is considered to be within the airspace directly below that area.

§91.107   Special VFR weather minimums.

(a) Except as provided in §93.113, when a person has received an appropriate ATC clearance, the special weather minimums of this section instead of those contained in §91.105 apply to the operation of an aircraft by that person in a control zone under VFR.

(b) No person may operate an aircraft in a control zone under VFR except clear of clouds.

(c) No person may operate an aircraft (other than a helicopter) in a control zone under VFR unless flight visibility is at least one statute mile.

(d) No person may take off or land an aircraft (other than a helicopter) at any airport in a control zone under VFR—

(1) Unless ground visibility at that airport is at least one statute mile; or

(2) If ground visibility is not reported at that airport, unless flight visibility during landing or takeoff is at least one statute mile.

§91.109   VFR cruising altitude or flight level.

Except while holding in a holding pattern of two minutes or less, or while turning, each person operating an aircraft under VFR in level cruising flight, at or above 3,000 feet above the surface, shall maintain the appropriate altitude prescribed below:

(a) When operating below 18,000 feet MSL and—

(1) On a magnetic course of zero degrees through 179 degrees, any odd thousand foot MSL altitude +500 feet (such as 3,500, 5,500, or 7,500); or

(2) On a magnetic course of 180 degrees through 359 degrees, any even thousand foot MSL altitude +500 feet (such as 4,500, 6,500, or 8,500).

(b) When operating above 18,000 feet MSL to flight level 290 (inclusive), and—

(1) On a magnetic course of zero degrees through 179 degrees, any odd flight level +500 feet (such as 195, 215, or 235); or

(2) On a magnetic course of 180 degrees through 359 degrees, any even flight level +500 feet (such as 185, 205, or 225).

(c) When operating above flight level 290 and—

(1) On a magnetic course of zero degrees through 179 degrees, any flight level, at 4,000-foot intervals, beginning at and including flight level 300 (such as flight level 300, 340, or 380); or

(2) On a magnetic course of 180 degrees through 359 degrees, any flight level, at 4,000-foot intervals, beginning at and including flight level 320 (such as fight level 320, 360, or 400).

INSTRUMENT FLIGHT RULES

§91.115 ATC clearance and flight plan required.

No person may operate an aircraft in controlled airspace under IFR unless—

(a) He has filed an IFR flight plan; and

(b) He has received an appropriate ATC clearance.

§91.116 Takeoff and landing under IFR: general.

(a) *Instrument approaches to civil airports.* Unless otherwise authorized by the Administrator (including ATC), each person operating an aircraft shall, when an instrument letdown to an airport is necessary, use a standard instrument approach procedure prescribed for that airport in Part 97 of this chapter.

(b) *Landing minimums.* Unless otherwise authorized by the Administrator, no person operating an aircraft (except a military aircraft of the United States) may land that aircraft using a standard instrument approach procedure prescribed in Part 97 of this chapter unless the visibility is at or above the landing minimum prescribed in that Part for the procedure used. If the landing minimum in a standard instrument approach procedure prescribed in Part 97 is stated in terms of ceiling and visibility, the visibility minimum applies. However, the ceiling minimum shall be added to the field elevation and that value observed as the MDA or DH, as appropriate to the procedure being executed.

(c) *Civil airport takeoff minimums.* Unless otherwise authorized by the Administrator, no person operating an aircraft under Part 121, [123,] 129, or 135 of this chapter may take off from a civil airport under IFR unless weather conditions are at or above the weather minimums for IFR takeoff prescribed for that airport in Part 97 of this chapter. If takeoff minimums are not prescribed in Part 97 of this chapter, for a particular airport, the following minimums apply to takeoffs under IFR for aircraft operating under those parts:

(1) Aircraft having two engines or less: 1 statute mile visibility.

(2) Aircraft having more than two engines: ½ statute mile visibility.

(d) *Military airports.* Unless otherwise prescribed by the Administrator, each person operating a civil aircraft under IFR into, or out of, a military airport shall comply with the instrument approach procedures and the takeoff and landing minimums prescribed by the military authority having jurisdiction on that airport.

(e) *Comparable values of RVR and ground visibility.*

(1) If RVR minimums for takeoff or landing are prescribed in an instrument approach procedure, but RVR is not reported for the runway of intended operation, the RVR minimum shall be converted to ground visibility in accordance with the table in subparagraph (2) of this paragraph and observed as the applicable visibility minimum for takeoff or landing on that runway.

(2) RVR

	VISIBILITY (STATUTE MILES)
1600 feet	1/4 mile
2400 feet	1/2 mile
3200 feet	5/8 mile
4000 feet	3/4 mile
4500 feet	7/8 mile
5000 feet	1 mile
6000 feet	1 1/4 mile

(f) *Use of radar in instrument approach procedures.* When radar is approved at certain locations for ATC purposes, it may be used not only for surveillance and precision radar approaches, as applicable, but also may be used in conjunction with instrument approach procedures predicated on other types of radio navigational aids. Radar vectors may be authorized to provide course guidance through the segments of an approach procedure to the final approach fix or position. Upon reaching the final approach fix, or position, the pilot will either complete his instrument approach in accordance with the procedure approved for the facility, or will continue a surveillance or precision radar approach to a landing.

(g) *Use of low or medium frequency simultaneous radio ranges for ADF procedures.* Low frequency or medium frequency simultaneous radio ranges may be used as an ADF instrument approach aid if an ADF procedure for the airport concerned is prescribed by the Administrator, or if an approach is conducted using the same courses and altitudes for the ADF approach as those specified in the approved range procedure.

(h) *Limitations on procedure turns.* In the case of a radar initial approach to a final approach fix or position, or a timed approach from a holding fix, or where the procedure specifies "NOPT" or "FINAL," no pilot may make a procedure turn unless, when he receives his final approach clearance, he so advises ATC.

APPENDIX 3

FAA SAMPLE EXAMINATION QUESTIONS ON WEATHER

In addition to the FAA regulations concerning weather, quoted above, the following are eleven sample questions and problems which are typical of those you will be required to answer for your FAA examination on weather:

(1) For VFR flight in a control area, the minimum flight visibility and proximity to cloud requirements are (choose the correct answer):

    1—Visibility 3 miles; 500 ft under, 1000 ft over, and 2000 ft horizontally from the clouds.

    2—Visibility 3 miles; clear of the clouds.

    3—Visibility 1 mile; 500 ft under, 1000 ft over, 2000 ft horizontally from the clouds.

    4—Visibility 1 mile; clear of the clouds.

(2) The Area Forecast for the period 7 A.M. to 7 P.M. CST(on page 379, indicates that the front will move eastward as a cold front. Should a squall line precede the front, it will normally be characterized by:

    1—fog, low stratus clouds, and stable air.

    2—hail, fog, and freezing precipitation.

    3—cold surface temperatures and stratus clouds.

    4—cumulus-type clouds, turbulence, and precipitation.

(3) Your attention is naturally attracted to the In-Flight Advisories, and the PIREPs, on page 380. Comparing the AirMet ALPHA 2 with the PIREP from SPS you conclude that:

1—The PIREP concerns a flight conducted well above the altitudes designated by the AirMet.

2—neither will be effective at your proposed takeoff time of 0830.

3—the turbulence which was forecast in the AirMet does not affect the area west of Wichita Falls.

4—neither pertains to your proposed route of flight.

(4) The Terminal Forecast on page 380 predicts a clear sky for Lubbock at 0800C. What visibility is forecast for 0800C?

1—15 miles.

2—6 miles.

3—10 miles.

4—Over 8 miles.

(5) Referring to the Winds Aloft Forecast on page 381, you estimate the winds aloft for 0800C at 5000 ft MSL for Lubbock to be from 280° at approximately 6 mph, and at 10,000 MSL to be from:

1—330° at 14 mph.

2—310° at 14 mph.

3—330° at 12 mph.

4—331° at 2 mph.

(6) A check of the 7:00 A.M. Aviation Weather Report for Wichita Falls, on page 381, indicates that:

1—the ceiling was 8000 ft.

2—an overcast ceiling existed.

3—the visibility was 8 miles.

4—there were scattered clouds at 8000 ft and a high thin overcast but no ceiling existed.

(7) You plan to monitor the voice feature of the Guthrie VOR enroute to Kickapoo to keep advised of the latest weather. Scheduled weather broadcasts will be available from Guthrie:

1—every 30 minutes at 15 and 45 minutes after the hour.

2—every hour at 30 minutes after the hour.

3—on the hour and on the half-hour.

4—every hour on the hour.

(8) Based on the 1300C Aviation Weather Reports on page 381, the surface wind at Mineral Wells should be from approximately:

1—080° at 12 knots.

2—350° at 8 knots and gusty.

3—350° at 8 mph and gusty.

4—300° at 12 mph and gusty.

(9) You may encounter areas of stratus clouds on this flight and you visualize the possibility of carburetor icing. Since your airplane is equipped with a fixed-pitch propeller, you realize that the indication of carburetor icing would likely be:

1—a decrease in engine RPM only.

2—engine roughness only.

3—a loss of power only.

4—any of the above.

(10) If you determine that carburetor icing does exist, which of the following methods would constitute the best *immediate* procedure?

1—Apply full "hot" carburetor heat to remove the existing ice and then follow the procedure as recommended by the manufacturer.

2—Climb or descend to another cruising level.

3—Move the carburetor heat control toward the full "hot" position until you get the maximum RPM increase.

4—Alternately move the carburetor heat control from the full "cold" position to the full "hot" position until you're sure the ice has removed.

(11) Light aircraft are particularly susceptible to wing tip vortices or wake turbulence. The most severe wake turbulence is produced by:

1—small aircraft during takeoff or landing.

2—large aircraft during takeoff or landing.

3—small aircraft in cruising flight.

4—large aircraft in cruising flight.

The answers to the above questions are as follows:

(1)   1. §91.105 states in part:

"(a) Distance from clouds. Except as provided in §91.107, no person may operate an aircraft under VFR

*   *   *   *   *   *

(2) Within any other controlled airspace (excludes continental control area) at a distance less than 500 feet below or 1,000 feet above, and 2,000 feet horizontally from, any cloud formation.

*   *   *   *   *   *

"(b) Flight visibility. Except as provided in §91.107, no person may operate an aircraft under VFR

*   *   *   *   *   *

(2) In any other controlled airspace (excludes the continental control area) unless flight visibility is at least 3 statute miles. . . ."

(2)   4. In some cases, an almost continuous line of thunderstorms may form along the front (cold) or ahead of it. These lines of thunderstorms, *squall lines,* contain some of the most turbulent weather experienced by pilots. Fog and stratus-type clouds (as indicated by answers 1, 2, and 3) are generally associated with a warm front.

(3)   3. Based on the weather forecast and Aviation Weather Reports available, it is doubtful that the moderate turbulence forecast by Airmet ALPHA 2 would exist as far west as Wichita Falls. With the additional evidence of only light turbulence, as

presented by the PIREP, you can safely conclude that the forecast moderate turbulence does not exist in the area west of Wichita Falls.

(4) 4. Referring to the Key to Aviation Weather Reports on page 383, you note that the absence of a visibility figure in Terminal Forecasts indicates that the visibility is over eight miles. (The Key to Aviation Weather Reports is always furnished with the Private Pilot Written Examination.) If you mistook the wind speed in the figure 2710 for visibility, you may have chosen incorrect response number 3.

(5) 1. The figure representing the Lubbock 10,000-ft wind at 0800 is 3312. This indicates a direction of 330° at 12 knots. Twelve knots is approximately 14 mph. If you had mistakenly interpreted the figure 12 as miles per hour, you would have chosen incorrect response number 3. (The Key to Aviation Weather Reports also includes an explanation of the Winds Aloft Forecast.)

(6) 4. The 0700C Aviation Weather Report for Wichita Falls shows a sky cover of 8000 ft scattered high, thin overcast, and a visibility of more than 15 miles. The height above ground of the lowest layer of clouds reported as broken or overcast and not classified as "thin" is the ceiling.

(7) 1. Every 30 minutes at 15 and 45 minutes after the hour. (In Canada, weather broadcasts start at 25 and 55 minutes past the hour.)

(8) 2. 3508 is properly interpreted only by response 2. Response 3 is incorrect since the wind is always reported in knots.

(9) 4. For airplanes with fixed-pitch propellers, the first indication of carburetor icing is loss of rpm. A roughness in engine operation may develop later. Concurrent with loss of rpm is loss of power.

(10) 1. A change in altitude might be a recommended procedure, in time, but it would not constitute the best *immediate* procedure. When ice is present, the full "hot" position should be used as the first step; not partial or alternating "hot" positions.

(11) 2. The most severe wake turbulence is produced by large aircraft in landing or takeoff configuration. Light aircraft are especially affected if they should encounter this type of turbulence. The heavier and slower the aircraft, the greater the intensity of the air circulation in the vortex cores. Therefore, responses 1, 3, and 4 are incorrect. Since vortices are not formed until lift is produced, they will not be generated on a takeoff roll until just before liftoff, or by a landing aircraft after it is solidly on the ground. Vortices settle downward and spread laterally. When it is necessary to operate

behind a large aircraft, try to remain above the flight path of that aircraft.

SAMPLE EXAMINATION

Plain Language Interpretation of the Area Forecast
For the Period 7 A.M. to 7 P.M., CST
For northwestern, north central, and northeastern Texas and Oklahoma

CLOUDS AND WEATHER: The stationary front which was located along the Oklahoma City—San Antonio line at midnight is now moving eastward as a weak cold front. Thunderstorms will occur along and ahead of the front and move eastward with the front. In the early forenoon, thunderstorms and rain showers will be located generally along a line from Oklahoma City to Fort Worth to Waco. By mid-afternoon, the area of thunderstorms should have moved to the extreme eastern portion of the area. Ceilings in the thunderstorm areas may be as low as 500 feet, with visibilities of $\frac{1}{2}$ to 1 mile due to rain and fog. Following frontal passage, ceilings should be generally unlimited and visibilities good. Ahead of the frontal system, surface winds should be southeasterly and gusty, to 25 knots. Behind the front, winds should be westerly to northwesterly, except southerly along the Texas-Oklahoma Border, at approximately 10 to 15 knots.

ICING: There will be locally moderate icing in clouds above the freezing level. The freezing level will be 13,000 to 15,000 feet.

TURBULENCE: Turbulence will be moderate to severe in the vicinity of thunderstorms. Behind the front turbulence should be light.

OUTLOOK: For the period 7 P.M. today until 7 A.M. tomorrow, the front will continue to move eastward and will be beyond this area by midnight.

Station Designators in Weather Reports Below

LBB — Lubbock
ABI — Abilene
MWL — Mineral Wells
FTW — Fort Worth
GSW — Greater Southwest International Airport
SPS — Witchita Falls

In-flight Weather Advisories

FL GSW 1811ØØ (5 A.M.)
SIGMET ALFA 1. OVR N CNTRL AND NERN TEX ALG AND E OF CLD FRNT MOD TO SVR TURBC IN TSTMS WITH HAIL TO 1 INCH DIAM. TSTMS FRMG PSBLY SLD LNS.

(Interpretation of Above SIGMET)

Over north central and northeastern Texas along and east of cold front,, moderate 'to severe turbulence can be expected in thunderstorms with hail to 1 inch in diameter. Thunderstorms forming possibly in solid lines.

FL GSW 1813ØØØ (7 A.M.)
AIRMET ALPHA 2. IMDT FLWG TO 50 MI W OF CLD FRNT IN N CNTRL TEX
OCNL MOD TURBC BLO 6Ø DUE TO RTHR STG GUSTY WNDS. LGT RN
SHWRS MAY RMN BRFLY AFT FROPA.

(Interpretation of Above AIRMET—Formerly
Designated ADVISORIES FOR LIGHT AIRCRAFT)

Immediately following to 50 miles west of cold front in north central Texas,
occasional moderate turbulence will exist below 6,000 feet due to strong,
gusty surface winds. Rain showers may remain briefly after frontal passage.

Pilot Reports

SPS PIREP 0738 GTH-SPS BLO SCTD CLDS 55 LGT TURBC. BE35

(Interpretation of SPS PIREP)

Originating station Wichita Falls. Time 7:38 A.M. CST. From Guthrie to
Wichita Falls below scattered clouds at 5,500 feet MSL light turbulence reported by a Beech Bonanza.

GSW PIREP 0810 SPS-V61 GSW INCLR AT 115 TIL BPR ⊕V⊕ HVY RN MOD
TURBC TO GSW. DC-6

(Interpretation of GSW PIREP)

Originating station Greater Southwest. Time 8:10 A.M. CST. From Wichita
Falls to Greater Southwest Airport via Victor 61 Bridgeport direct Greater
Southwest at 11,500 feet MSL clear until Bridgeport. Broken clouds variable
to overcast for remainder of flight with heavy rain and moderate turbulence
reported by a DC-6.

Terminal Forecasts

Period 0500C–1700C (5 A.M.–5 P.M. CST)

FT1 181045
11Z–23Z MON
LBB 5Ø⊕7. 0800C O 2710
ABI C35⊕6Ø⊕7 2415. 1000C 8ØⒺ 2612
GSW 1Ø⊕2TRW 1320G25 VRBL 1620G25. 1100C FROPA 4ØⒺC8ØⒺ7 OCNL
R—2710G15. 1300C 9ØⒺ 3010G15.

Aviation Weather Reports

(Teletype Sequence Reports)

0700C
Ø29 SA291813ØØ CIRCUIT 8029, 18th DAY OF MONTH, 1300 GREENWICH
TIME (Z) OR 0700 CENTRAL STANDARD TIME (C)
LBB 6ØⒺ7 129/54/46/25Ø8/997
ABI E8Ø⊕15 133/59/52/34Ø4/997
MWL S E5⊕1 TRWF 128/64/63/3212G18/993/TB06 OVHD MOVG EWD LTGIC
ALQDS
FTW M13⊕35⊕1Ø 127/7Ø/68/1212G18/991/⊕V⊕

GSW M11⊕7 127/71/66/1315G24/993/RB27 RE43
Ø3Ø (CIRCUIT 8030)
SPS 8ØⓄ/-⊕15+ 132/81/61/1922G27
0800C
Ø29 SA291814ØØ
LBB O15+ 132/57/42/27Ⓞ8/998/FEW CU SE
ABI E9ØⓊ15 133/61/5Ø/29Ø8/997/CLDS DRK SE
MWL M25Ⓤ/Ⓤ1Ø 129/65/59/3Ø12G18/995/TSTM SE MOVG E
FTW M2ØⓊ5Ø⊕1Ø 128/72/65/221ØG16/993/RB34 RE5Ø TSTM E MOVG SE
CLDS DRK SW WSHFT 0740C
GSW M2Ø⊕7TRW- 127/7Ø/68/162ØG26/993/LN TSTMS SW-NE MOVG E
Ø3Ø
SPS /Ⓤ15+ 135/8Ø/55/1715G2Ø/998
1300C
Ø29 SA291819ØØ
LBB O25 134/61/4Ø/271Ø/999
ABL O15+ 134/64/46/281Ø/999
MWL E9ØⓊ15 133/66/5Ø/35Ø8G12/998
FTW E1ØØⓊ15 131/68/52/29Ø6/997
GSW O12 13Ø/69/5Ø/311ØG15/996
Ø3Ø
SPS O2Ø 135/78/5Ø/181Ø/998

Winds Aloft Forecasts
0600C—1200C
FD WBC 181150
12-18Z

LVL	3000	5000 FT	7000	10000 FT	15000 FT	20000 FT	25000 FT
LBB		2805	3112	3312+04	3513−06	3517−17	3519−28
GSW	9900	9900+10	3605	3312+06	3316−05	3312−15	3322−25

1200C–1800C

18-24Z

LVL	3000	5000 FT	7000	10000 FT	15000 FT	20000 FT	25000 FT
LBB		2810	3010	3110+04	3412−05	3416−16	3420−26
GSW	3408	3310+08	3112	3012+05	3016−05	3018−15	3020−25

("9900" is used to indicate winds of less than 5 knots. It is spoken of as "light and variable.")

KEY TO AVIATION WEATHER REPORTS . . . .

Location Identifiers	Sky and Special Report +	Visibility Weather and Obstruction to Vision	Ceiling	Sea Level Pressure	Temperature and Dew Point	Wind	Altimeter Setting	Runway Visual Range	Coded Pireps	Remarks
MKC	S 15⊙ M25⊕	4R –K	132	/58/56		/1807	/993	/VR32	/⊕55	RB05V⊕

SKY AND CEILING

Sky cover symbols are in ascending order. Figures preceding symbols are heights in hundreds of feet above station.

Sky cover Symbols are:

O=Clear: Less than 0.1 sky cover.
⊙=Scattered: 0.1 less than 0.6 sky cover.
⊕=Broken: 0.6 to 0.9 sky cover.
⊕=Overcast: More than 0.9 sky cover.
–=Thin (When prefixed to the above symbols.)
-X=Partial Obscuration: 0.1 to less than 1.0 sky hidden by precipitation or obstruction to vision (bases at surface).
X=Obscuration: 1.0 sky hidden by precipitation or obstruction to vision (bases at surface).

Letter preceding height of layer identifies ceiling layer and indicates how ceiling height was obtained. Thus:

A Aircraft	M Measured	/ Height of cirriform
B Balloon (Pilot or ceiling)	R Radiosonde Balloon or Radar	nonceiling clouds nonceiling layer unknown
D Estimated height of cirriform clouds on basis of persistency	W Indefinite	"v" immediately following numerical value indicates a varying ceiling
E Estimated	U Height of cirriform ceiling layer unknown	

VISIBILITY

Reported in Statute Miles and Fractions.
(V = Variable)

WEATHER SYMBOLS

A=Hail	R=Rain	SW=Snow Showers
AP=Small Hail	RW=Rain Showers	T=Thunderstorm
E=Sleet	S=Snow	ZL=Freezing Drizzle
EW=Sleet Showers	SG=Snow Grains	ZR=Freezing Rain
IC=Ice Crystals	SP=Snow Pellets	
L=Drizzle		

INTENSITIES are indicated thus:
-- Very light — Light (no sign)
Moderate + Heavy

OBSTRUCTION TO VISION SYMBOLS

D=Dust	H=Haze	Dust
F=Fog	IF=Ice Fog	BN=Blowing Sand
GF=Ground Fog	K=Smoke	BS=Blowing Snow
	BD=Blowing	

WIND

Direction in tens of degrees from true north, speed in knots. 0000 indicates calm. G indicates gusty. Peak speed of gusts follows G or Q when squall is reported. The contraction WSHFT followed by local time group in remarks indicates windshift and its time of occurrence.

EXAMPLES: 3627   360 Degrees, 27 Knots.
0127   010 Degrees, 27 Knots.
1027   100 Degrees, 27 Knots.
3627G40   360 Degrees, 27 Knots Peak speed in gusts 40 Knots.

ALTIMETER SETTING

The first figure of the actual altimeter setting is always omitted from the report.

RUNWAY VISUAL RANGE (RVR)

RVR is reported only from selected stations. The value reported is a 10-minute mean of the visual range in hundreds of feet.

CODED PIREPS

Pilot reports of clouds not visible from ground are coded with NSL height date preceding and/or following sky cover symbol to indicate cloud bases and/or tops, respectively.

DECODED REPORT

Kansas City. Special observation, 1500 feet scattered clouds, measured ceiling 2500 feet overcast, visibility 4 miles, light rain, smoke, sea level pressure 1013.2 millibars, temperature 58°F, dewpoint 56°F, wind 180°, 7 knots, altimeter setting 29.93 inches. Runway Visual Range 3200 feet, pilot reports top of overcast 5500 feet, rain began 5 minutes past the hour, overcast variable broken. + S indicates that report contains important change.

TERMINAL FORECASTS contain information for specific airports on ceiling, cloud heights, cloud amounts, visibility, weather condition and surface wind. They are written in a form similar to the AVIATION WEATHER REPORT.

CEILING: Identified by the letter "C"

CLOUD HEIGHTS: In hundreds of feet above the station

CLOUD LAYERS: Stated in ascending order of height

VISIBILITY: In statute miles, but omitted if over 8 miles

SURFACE WIND: In tens of degrees and knots; omitted when less than 10.

Examples of TERMINAL FORECASTS:

C15⊕   Ceiling 1500', broken clouds.

C15⊕6K   Ceiling 1500' overcast, visibility 6 miles, smoke.

20⊕C70⊕3230G Scattered clouds at 2000', ceiling 7000' overcast, surface wind 320 degrees 30 knots, gusty

O1-1/2GF   Clear, visibility one and one-half miles, ground fog.

C5x1/4S   Sky obscured, vertical visibility 500', visibility one fourth mile, moderate snow.

AREA FORECASTS are 12-hour forecasts of cloud and weather conditions, cloud tops, fronts, icing and turbulence for an area the size of several states. A 12-hour OUTLOOK is added. Heights of cloud tops, icing, and turbulence are above SEA LEVEL.

SIGMET advisories include weather phenomena potentially hazardous to all aircraft.

AIRMETs include weather phenomena of less severity than that covered by SIGMETs which are potentially hazardous to aircraft having limited capability due to lack of equipment or instrumentation or pilot qualifications and are at least of operational interest to all aircraft.

WINDS ALOFT FORECASTS provide a 12-hour forecast of wind conditions at selected flight levels. Temperatures will be included to all levels above 3000, except 5000 feet when this is the lowest level forecast and 7000.

EXAMPLE:

LVL 3000 5000FT 7000 10000FT 15000FT 20000FT 25000FT

MKC 2222 2220+19 2220 2414+10 2713+01 2815—09 2815—17

3000FT (MSL) 220° 22KT

10000FT (MSL) 240° 14KT TEMP +10°C.

PILOTS report in-flight weather to nearest FSS.

APPENDIX 4

MINIMUM INFORMATION FOR FLIGHT PLANNING, SELF-BRIEF

a. Departure Conditions
    (1) Bases and Tops of Clouds
        (a) Existing from:
            Local observation
            Pilot reports
            Local radar
        (b) Forecast from:
            Local forecast
            Terminal forecasts
            Area forecasts
            Twelve-hour significant weather chart
            High-level significant weather chart
    (2) Visibility and Precipitation
        (a) Existing from:
            Local observation
            Sequence report
        (b) Forecast from:
            Local or terminal forecast
    (3) Freezing Level (0° to 18°C.)
        (a) Existing from:
            Freezing-level chart
            Comparison of constant-height charts

      (b) Forecast from:
         Area forecast
   (4) Temperature and Winds to Flight Altitude
      (a) Average winds from:
         Winds aloft charts
         Constant-pressure charts
         Winds aloft codes
      (b) Temperature from:
         Constant-pressure chart
   (5) Runway Temperature and Pressure Altitude
   From local forecast's graph or current observation adjusted for diurnal change

b. En Route Weather
   (1) Bases, tops, type, and amount of each cloud layer
      (a) Existing from:
         Surface chart
         Sequence reports
         Pilot reports
         Radar summary
      (b) Forecast from:
         Terminal forecast
         Area forecast
         Twelve-hour significant weather chart
         High-level significant weather prog
         Twenty-four-hour surface prog
   (2) Visibility
      (a) Surface:
         Sequence
         Surface chart
         Terminal forecast
      (b) Aloft:
         High-level cloud types taken from high-level significant weather prog
   (3) Type, Location, Intensity, Direction, and Speed of Frontal Movement
      (a) Surface chart: type and location
      (b) Radar summary: intensity
      (c) Area and terminal forecasts
      (d) Twenty-four-hour surface prog (frontal direction and speed)
      (e) Twelve-hour significant weather chart
   (4) Freezing level from:
      (a) Freezing-level chart
      (b) Constant-pressure chart
      (c) Area forecast
   (5) Temperature and Winds at Flight Altitude
      (a) Winds aloft charts
      (b) Constant-pressure charts

(c) Upper-wind forecast (teletype)
(6) Areas of Severe Weather (thunderstorms, hail, icing, and turbulence)
   (a) Surface, stability index, local radar, and radar summary charts
   (b) Pilot reports
   (c) Twelve-hour significant weather chart
   (d) High-level significant weather prog chart
   (e) Area forecast

c. Destination and Alternate
(1) Bases, Tops, and Amount of Cloud Layers
   (a) Existing from:
      Surface airways reports (sequences)
      Surface chart
      Pilot reports
   (b) Forecast from:
      Terminal forecast
      Twelve-hour significant weather prog chart
      High-level significant weather chart
(2) Visibility
   (a) Existing from:
      Sequence report (surface airways)
   (b) Forecast from:
      Terminal forecast
(3) Precipitation (type and intensity)
   (a) Existing from:
      Sequence reports (surface airways)
   (b) Forecast from:
      Terminal forecast
(4) Freezing Level from:
   (a) Freezing-level chart
   (b) Area forecasts
   (c) Comparison of constant-pressure charts.
(5) Surface Wind Speed and Direction
   (a) Existing from:
      Sequence reports (surface airways)
   (b) Forecast from:
      Terminal forecast
(6) Forecast Altimeter Setting from:
   (a) A local forecaster
   (b) Possible by adjusting forecast of minimum altimeter setting (QNH) in terminal forecast with tendency from surface chart and/or sequence report

APPENDIX 5

WEATHER EXAMINATION FOR
STUDENT MILITARY PILOTS

1. You are planning an airways flight from Oklahoma City (OKC), via FSM, LIT, GRW, JAN, MEI, EVR, and DHN, to Tallahassee (TLH). Your departure time is 1030Z and your ETA is 1500Z on 7 November. DHN is planned alternate and 1524Z is the alternate ETA.
2. Sequence reports:
   030 SA30071000
   OKC M14⊕8 168/38/31/3417/003
   027 SA27071000
   FSM M15⊕5L− 140/42/37/3315/993
   LIT M17⊕10 113/45/40/3320/986
   028 SA28071000
   GRW 5⊕W18⊕4RW 068/61/60/2017G28/973
   JAN 4⊕M20⊕3TRW 071/63/61/2023G39/974
   MEI E25⊕6 089/66/62/1818G34/979/LTNG W
   EVR M12⊕3L− 124/65/62/1815/990
   DHN M11⊕6 140/65/63/1814/995
   024 SA24071000
   TLH M10⊕3F 159/63/61/1612/300
3. Terminal Forecasts:
   FT1 GSW 070445
   05Z − 17Z MON
   OKC O 3320. 08Z 15⊕ 3320. 09Z C15⊕ 3315.
   FT1 MEM 070445
   05Z − 17Z MON
   FSM C20⊕ 3320. 09Z C15⊕ 3315.
   LIT C20⊕ 3320. 09Z C15⊕ 3315.

389

GRW 5⊙C20⊕3TRW 2120G. 11Z COLD FROPA
C5⊕2TRW 3130G. 13Z 5⊙C20⊕ 3220. 14Z
C20⊕ 3215. .
FT1 NEW 070445
05Z – 17Z SUN
JAN 5⊙C20⊕3TRW 2120G. 09Z C3⊕1TRW SVR
SQUALL LINE 2035G OCNL LWRG C1X1/2TRW PSBL
HAIL. 15Z COLD FROPA C5⊕3RW 2830G. .
MEI C20⊕4 1820. 08Z 5⊙C20⊕3TRW 2025G
OCNL C5X1TRW. 14Z C3⊕1TRW SVR SQUALL
LINE OCNL LWRG C1X1/2TRW PSBL HAIL. .
EVR C12⊕3L– 1815. 11Z C10⊕2L–F 1815. 13Z
C8⊕1L–F 1815. 15Z C9⊕3L–F 1815. 16Z
C10⊕5H 1817. .
FT2 JAX 070520
05Z MON – 05Z TUE
DHN C12⊕6 1615. 09Z C10⊕3L–F 1615. 12Z
C8⊕2L–F 1613. 14Z C9⊕3F 1613. 16Z C10⊕4F. 1613.
18Z C15⊙200⊙6 1615. 20Z 20⊙200⊙ 1620. 23Z 200⊙ 1820.
03Z C20⊕ 1820 OCNL RW. .
TLH C10⊕3L–F 1613. 09Z C8⊕2L–F 1613. 11Z
C7⊕1L–F. 1330Z C8⊕3F. 1430Z C10⊕5 1610. 16Z
C12⊕6 1615. 18Z C20⊙200⊙ 1615. 20Z 20⊙200⊙ 1620.
23Z 200⊙ 1820. .

4. Published minimums for TLH:
   a. ADF: 1,000'–1 mile
   b. VOR: 500'–1 mile
   c. PAR: 250'–½ mile
5. Published minimums for DHN:
   a. ADF: 800'–2 miles
   b. VOR: 700'–1 mile
   c. TACAN: 500'–1 mile

METEOROLOGY EXAMINATION QUESTIONS

(1) What type of frontal system would you encounter en route? (See Figure 1.)
(2) Which side of the frontal system has the colder air? (See Figure 1.)
(3) Is the frontal system an abrupt or shallow type of system? (See Figures 1 through 5.)
(4) In what direction is the cold wedge of air expected to move? (See Figures 1 and 6.)
(5) Since thunderstorm activity is expected en route, what intensity of turbulence should be expected in the thunderstorm area? (See Figure 7.)
(6) To what altitude should the thunderstorm activity extend along the line of flight? (See Figure 7.)

(7) Between what altitudes in the thunderstorm activity can maximum turbulence be expected? (See Figure 7.)

(8) Between what altitudes in the thunderstorm activity can structural icing be expected? (See Figure 7.)

(9) What are the altitudes of the freezing level en route? (See Figure 8.)

(10) What portion of the route is classified as a severe-weather warning area? (See Figure 9.)

(11) What portion of the line of flight has a ceiling of less than 1000 ft and/or a visibility of less than 3 miles? (See Figure 10.)

(12) If you had requested and received the Horizontal Weather Depiction Chart, at what altitude could you top the clouds over the State of Oklahoma? (See Figure 11.)

(13) What is the predicted top of the overcast over the destination Tallahasee? (See Figure 11.)

(14) What is the lowest *current* en route ceiling? (See Figure 12 and Sequence Report, pages 389–390.)

(15) What is the lowest *forecast* en route ceiling? (See Terminal Forecasts, pages 389–390.)

(16) Between what times is the front forecast to pass GRW? (See GRW Terminal Forecast page 390.)

(17) Between what times is the front forecast to pass JAN? (See JAN Terminal Forecast, page 390.)

(18) What are the forecast ceiling and visibility at TLH at ETA? (See TLH Terminal Forecast, page 390.)

(19) If you hold a standard instrument card, could you legally depart IFR? (See OKC Sequence Report, page 389.)

(20) Could you legally file VFR? (See Sequence Report and Terminal Forecasts, pages 389–390.)

(21) If you intend to make an ADF approach at TLH (ETA 1500Z), can you legally file IFR to TLH? (See TLH Terminal Forecast and Published Minimums, page 390.)

(22) If you intend to make a VOR approach at TLH, can you legally file IFR? (See TLH Terminal Forecast and Published Minimums, page 390.)

(23) If you intend to make a PAR approach at TLH, can you legally file IFR? (See TLH Terminal Forecast and Published Minimums, page 390.)

(24) Why is an alternate required for this flight? (See Terminal Forecasts, page 390, and Alternate Airport Rules.)

(25) Can you legally use DHN (ETA 1524Z) as an alternate if you intend to make an ADF approach at DHN? (See DHN Terminal Forecast and Published Minimums on page 390.)

(26) Can you legally use DHN as an alternate if you intend to make a VOR approach at DHN? (See DHN Terminal Forecast and Published Minimums, page 390.)

(27) Can you legally use DHN as an alternate if you intend to make a TACAN approach at DHN? (See DHN Terminal Forecast and Published Minimums, page 390.)

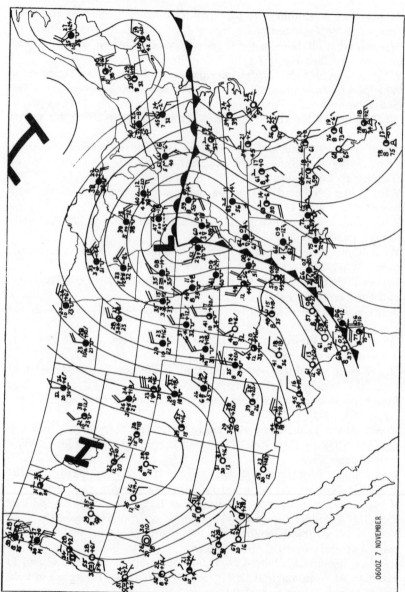

Figure 1 Surface (Synoptic) Weather Map.

0600Z 7 NOVEMBER

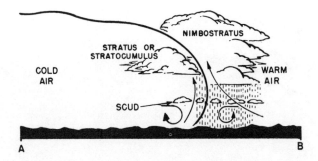

Figure 2 Horizontal and Vertical Cross Sections of an Abrupt-type Front with Stable Upgliding Air.

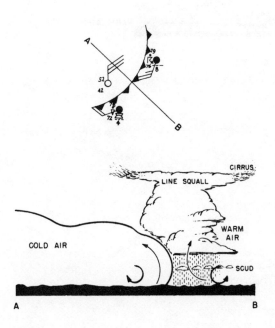

Figure 3 Horizontal and Vertical Cross Sections of an Abrupt-type Front with Conditionally Unstable Upgliding Air.

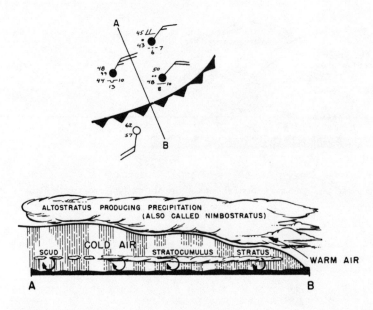

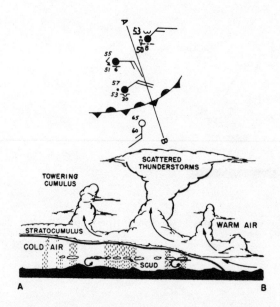

Figure 4 Horizontal and Vertical Cross Sections of a Shallow-type Front with Stable Upgliding Air.

Figure 5 Horizontal and Vertical Cross Sections of a Shallow-type Front with Conditionally Unstable Upgliding Air.

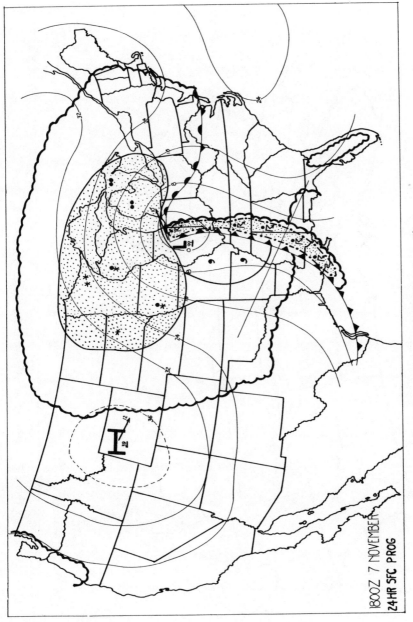

Figure 6 Twenty-four Hour Surface Prog Chart.

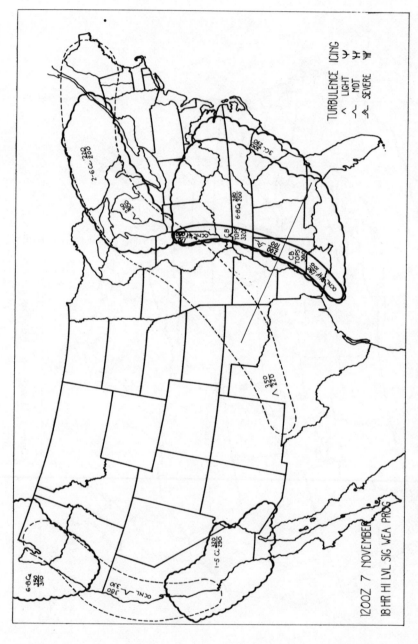

Figure 7 Eighteen-hour High-Level Significant Weather Prog Chart.

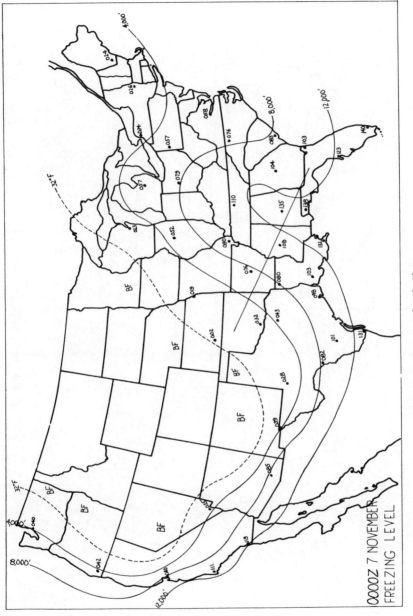

Figure 8 Freezing-level Chart.

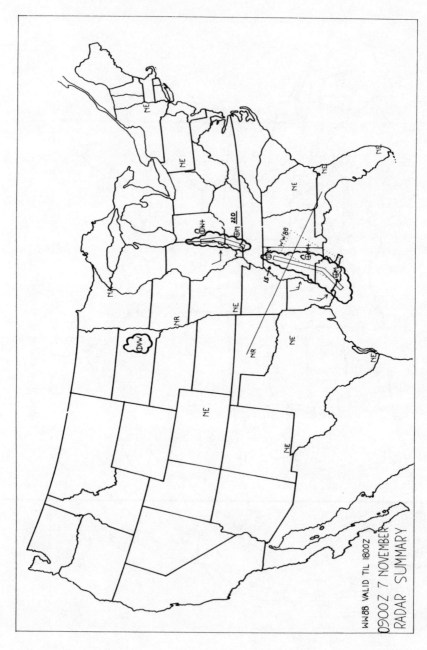

Figure 9 Radar Summary Chart.

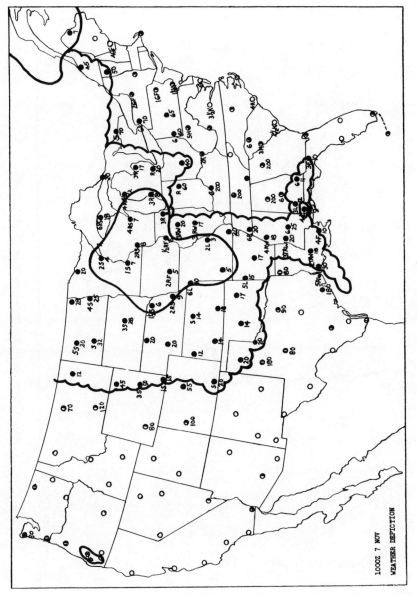

Figure 10   Weather Depiction Chart.

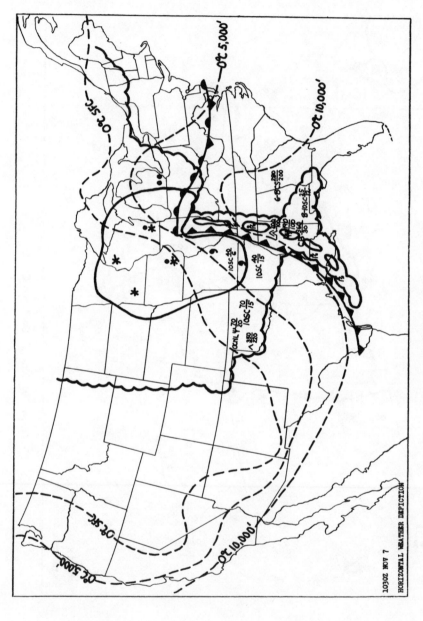

Figure 11 Horizontal Weather Depiction Chart.

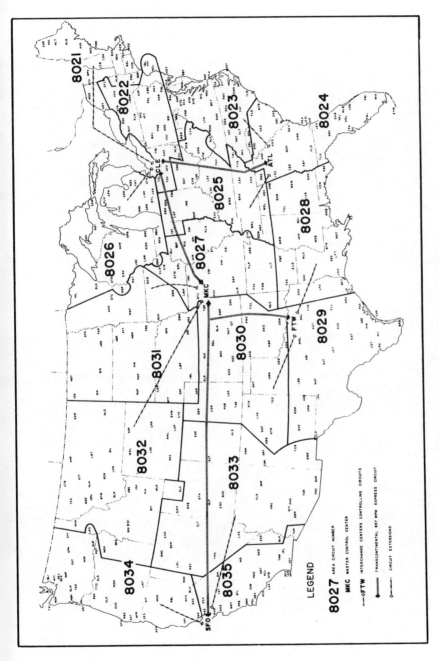

Figure 12 Weather Bureau Teletype Area Map.

INDEX

Page numbers in *italics* indicate figures.